ANDEAN ECOLOGY

Dellplain Latin American Studies

No. 8 *Studies in Spanish American Population History*, edited by David J. Robinson

No. 9 *The People of the Colca Valley: A Population Study*, Noble David Cook

No. 10 *Tierra Adentro: Settlement and Society in Colonial Durango*, Michael M. Swann

No. 11 *Andean Reflections: Letters from Carl O. Sauer While on a South American Trip Under a Grant from the Rockefeller Foundation, 1942*, Edited by Robert C. West

No. 12 *Credit and Socioeconomic Change in Colonial Mexico: Loans and Mortgages in Guadalajara, 1720 - 1820*, Linda Greenow

No. 13 *Once Beneath the Forest: Prehistoric Terracing in the Río Bec Region of the Maya Lowlands*, B. L. Turner, II

No. 14 *Marriage and Fertility in Chile: Demographic Turning Points in the Petorca Valley, 1840 - 1976*, Robert McCaa

No. 15 *The Spatial Organization of New Land Settlement in Latin America*, Jacob O. Maos

No. 16 *The Anglo-Argentine Connection, 1900 - 1939*, Roger Gravil

No. 17 *Costa Rica: A Geographical Interpretation in Historical Perspective*, Carolyn Hall

No. 18 *Household Economy and Urban Development: São Paulo, 1765 - 1836*, Elizabeth Anne Kuznesof

No. 19 *Irrigation in the Bajío Region of Colonial Mexico*, Michael E. Murphy

No. 20 *The Cost of Conquest: Indian Decline in Honduras Under Spanish Rule*, Linda Newson

No. 21 *Petty Capitalism in Spanish America: The Pulperos of Puebla, Mexico City, Caracas, and Buenos Aires*, Jay Kinsbruner

No. 22 *British Merchants and Chilean Development, 1851- 1886*, John Mayo

No. 23 *Hispanic Lands and Peoples: Selected Writings of James J. Parsons*, edited by William M. Denevan

No. 24 *Migrants in the Mexican North: Mobility, Economy, and Society in a Colonial World*, Michael M. Swann

No. 25 *Puebla de los Angeles, Industry and Society in a Mexican City, 1700 - 1850*, Guy P. C. Thomson

No. 26 *Generations of Settlers: Rural Households and Markets on the Costa Rican Frontier, 1850 - 1935*, Mario Samper

Dellplain Latin American Studies

PUBLISHED IN COOPERATION WITH
THE DEPARTMENT OF GEOGRAPHY
SYRACUSE UNIVERSITY

ANDEAN ECOLOGY

ADAPTIVE DYNAMICS IN ECUADOR

Gregory Knapp

Dellplain Latin American Studies, No. 27

Routledge
Taylor & Francis Group
NEW YORK AND LONDON

First published 1991 by Westview Press

Published 2018 by Routledge
605 Third Avenue, New York, NY 10017
2 Park Square, Milton Park, Abingdon, Oxon OX14 4RN

Routledge is an imprint of the Taylor & Francis Group, an informa business

A CIP catalog record for this book is available from the Library of Congress.

ISBN 13: 978-0-367-01632-6 (hbk)

Contents

Tables

List of Figures and Plates

FIGURES

PLATES

Preface and Acknowledgments

This book reflects a decade of research and travel in the equatorial Andes. I spent over 28 months in Ecuador from 1979 to 1987, travelling on foot, by jeep, on horseback, and by truck and bus in an effort to familiarize myself with the Ecuadorian highlands as a whole as well as with the various study sites. During this period, my initial ideas, prejudices, and enthusiasms have sometimes given way to different ideas and interpretations. In summarizing my research on traditional and prehistoric adaptive strategies, I have taken the opportunity not only to present new results but to reinterpret and update relevant earlier research.

The research on which this book is based was made possible by an Organization of American States grant for study in Ecuador (1979), a Fulbright Grant for Dissertation Research (1980-1981), a National Science Foundation Grant (144R345) in 1981, a Tinker Grant (1984), a Summer Research Award and Research Grant from the University Research Institute of the University of Texas (1985), and a Fulbright-Hays Faculty Research Abroad Grant (1986-1987).

In Ecuador, my work was facilitated by the Comisión Fulbright (Marcelo Erazo and Gonzalo Cartagenova, directors, with much help from María Mogollón and Jenny Castillo). The Instituto Nacional del Patrimonio Cultural (and especially María del Carmen Molestina, Ana V. de Jaramillo, Rodrigo Erazo, and Juan Carrera Colin) provided me with permits, advice, and field assistance in archaeology. The Museo del Banco Central also helped me with its facilities, and I particularly wish to thank its directors, Hernán Crespo Toral and María del Carmen Molestina, as well as the affiliated archaeologists, Ron Lippi and Antonio Fresco, who proved invaluable field companions. The Instituto Otavaleño de Antropología provided important facilities in the northern highlands; I especially thank José Echeverría for his invaluable assistance, field companionship, and hospitality. Ronaldo Mosquera A., head of the Ibarra branch of the Banco Central del Ecuador, provided help in gaining access to the archive there, and Mercedes Moreno Almeida generously gave of her time and knowledge while using that archive.

The Centro Panamericano de Estudios e Investigaciones Geográficas provided my initial training in Ecuadorian natural resources in 1979, and I enjoyed the advice and help of Roy Ryder, its academic director for many years. At the Pontífica Universidad Católica del Ecuador in Quito, I benefitted from the help of geographer Jaime Miranda and anthropologists Segundo Moreno Yánez and José Pereira. Carlos Cazco of the Programa Investigación en Producción helped me obtain and interpret their valuable research results on small farm production. I also benefitted from the generosity of the members of the Asociación Indígena Evangélica de Chimborazo, who helped make my work in western Chimborazo possible.

Many individuals in the Programa de Regionalización Agraria facilitated my search for agricultural information and helped in the interpretation of that information; I am especially indebted to Luis Cañadas, but also wish to thank Oscar Rovere, Luis Mejía, Washington Estrada, Trajano Yugcha, Tarquino Salcedo, Raul Bravo, and Freddy López. Pierre Gondard, Pierre Pourrut, and the other members of the ORSTOM technical assistance team were also extraordinarily helpful, especially so in their critiques of my ideas. Lenny Field, Manuel Chiriboga, José Sánchez Parga, and Galo Ramón of the Centro Andino de Acción Popular were very helpful, in discussing and helping in the publication of my research. In this regard I also owe a debt of thanks to José Echeverría of *Sarance*, Irving Zapoter and Ramiro Salvador of the Centro de Investigación y Cultura of the Central Bank of Ecuador, and María del Carmen Molestina, all of whom spent so much time in helping me clarify my ideas and getting them published in Spanish. Thanks to their help, many of the results published in English in this book have been made available to Ecuadorians in easily accessible publications.

Carlos Suárez of Ibarra, an outstanding field geographer and teacher, introduced me to the agriculture of the northern highlands. Patricia Mothes and Roy Jameson helped me study the canal systems of the northern highlands, and I also benefitted from field excursions with David Preston, John Isaacson, Kent Mathewson, John Treacy, James Parsons, William Denevan, and Martin Parry. I am grateful for Martin Parry's help and guidance in getting me involved with the Ecuadorian component of the Climate Impacts Project of the United Nations University, United Nations Environmental Program, and International Institute of Applied Systems Analysis.

Numerous other individuals helped me find important sources of information and warned me of pitfalls; among them, I am especially indebted to Alfredo Costales, Presley Norton, Jaime Borja of the Secretaría de Desarollo Rural Integral (SEDRI), Diego del Alcazar of the Instituto Nacional de Meteorología, Fausto Maldonado, Carlos Coba, Misael Acosta-Solís, Carlos Puga, Juan Botasso, Leslie Brownrigg, Barbara Schroeder, Philip Westra, Chantal Caillavet, Georges de Noni, Jorge Reyes, Bruce Kernan, Chalmers Clapperton, Ramón Vera, and Minard (Pete) Hall. In Peru I benefitted from conversations with Mario Tapia, Jorge Flores, Luis Lescano, Nicole Bernex de Falen, Hildegardo Córdova, Franklin Pease, María Rostworowski, and John Rowe, and from a field excursion with Clark and Kay Erickson.

Much of this research has required the help and hospitality of farmers. It is impossible to list them all, but I am especially indebted to the hospitality of the late Galo Plaza Lasso, who gave me free access to his haciendas and allowed me to study his account books for agronomic information. Other farmers who were of special help include Gustavo Hinojosa of San Pablo, Carlos Pérez of Hacienda La Vega, Federico Neumann, Miguel Ruíz, Manuel María López, José Villa, Luis Folleco, Genaro Villa, Luis Alfonso Rivera, Miguel Sotelo, Luis Valles, Amado Puga, Alberto Guevaro, Victor Valdevieso, Rigoberto Cruz, Sixto Espinosa, Luis Flores, José Miguel Quita, Anibal Sevilla Terán, Olmedo

Rigoberto Padilla, Vicente Tobar Zuleta, Emilio Bonifaz, Marcos Wray Espinosa, Francisco Narváez, Tom Gillespie, Luis Raza, Carlos Pilco, José María Yuquilema, Francisco Chimbolema Huacho, Manuel Cepeda Chicaisa, Francisco Bueno, Alberto Guillermo Cepeda Anilima, and Bolívar Troya.

The overall inspiration for much of my work is due to William Denevan, chair of my Ph.D. committee at Wisconsin. Other early influence and help came from Louisa Stark, Donald Thompson, Thomas Vale, Daniel Doeppers, Francis Hole, Frank Salomon, John Kutzbach, Bruce Batchelor, Alan Osborne, John Stephen Athens, Thomas Myers, and from my fellow graduate students Kent Mathewson, John Treacy, Ron Lippi, and Stuart White. Others who have reviewed or commented on portions or earlier incarnations of this book include Robert Netting, Emilio Moran, David Browman, and Bruce Winterhalder. Philip Porter, Diana Liverman, Phil Gersmehl, Stephen Brush, Dan Gade, Clark Erickson, B. L. Turner II, and William Mitchell have also been useful to me for their advice or criticism. I have benefitted from discussions with colleagues at Texas, including Richard Schaedel, Karl Butzer, William Doolittle, Terry Jordan, and Paul English.

I also would like to thank Sang Sook for her patience and the editor of this volume, David Robinson, for his useful comments and continued help.

Gregory Knapp

Chapter 1

INTRODUCTION

Andean Ecology

A remarkable assemblage of agricultural techniques, and an equally remarkable set of diverse soils have supported the changing societies of the Andean region. This book describes and analyzes the strategies of traditional and prehistoric farmers in one part of the Andes, in an effort to understand the varying interactions between people and their habitat over the last five hundred years.

Adaptive Strategies

Although this is a study of culture and environment in the Andes, it is not meant to be a regional cultural geography, a holistic analysis of a relatively stable or climax human ecosystem, or an application of social or world-systems theory to a peasant society. This is an examination and characterization of options, in terms which can be used to explore the environmental and technical conditions of cultural landscapes, cultural systems, and societies.

The aim has been to focus on those traits which can most reasonably be considered means rather than goals, and thus most subject to criteria of maximization of efficiency. These include such tactics or components of technology as fallowing, fertilization practices, and modifications of the soil surface (agricultural landforms).

These tactics can be studied on the contemporary scene, and documented in part for past periods using historical documentation and archaeological evidence.[1] The characterization of these traits in terms of their respective functions or efficiencies allows the construction of scenarios, predictions and "retrodiction" (predictions of past situations on theoretical principles).[2] Although full retrodiction or prediction is never possible, such attempts can serve as tests of our understanding of cultural goals and our knowledge of the technical characteristics of means to achieve these goals. Indeed, without some at least implicit attempt at modelling, certain aspects of prehistory (and history) must remain obscure. In practice, of course, prehistorians, like development experts, usually rely on assumptions of the possible, efficient, functional, and practical.

Cultures seldom have arrived at completely stable states. More commonly, the choice of efficient means to goals has been an ongoing process, disrupted by catastrophes, the consequences of other actors' actions, contact between groups, and change of the goals themselves. Choice of options can be considered part of an ongoing process of adaptation.[3]

A concern with adaptive process places emphasis on the identification of selective pressures, variability of cultural practices, and the cultural, social, and economic criteria of selection or choice between alternatives. In the Andes, there have been several technological revolutions connected with both cultural diffusion and change in socio-economic contexts. These historical processes of diffusion and institutionalization help explain the availability of options and the selective pressures facing folk decision-makers, but they do not explain why or how alternatives for action are in fact adopted. Here it is assumed that Andean subsistence farmers have always been aware of a variety of agricultural techniques, and that the actual techniques employed have usually been those which supplied needs and wants with the least expenditure of labor, given available soil and capital resources as conditioned by socially-determined access to resources.

It is not assumed that labor efficiency (or any other single factor) has governed cultural evolution as a whole, or that Andean peoples are subject to rigid laws of behavior. Rather, the aim has been to help understand the challenges and achievements of these mountain cultures in achieving

---

1 Karl W. Butzer, *Early Hydraulic Civilization in Egypt: A Study in Cultural Ecology*, Chicago: University of Chicago Press, 1976; Karl W. Butzer, *Archaeology as Human Ecology: Theory and Method for a Contextual Approach*, New York: Cambridge University Press, 1982; Karl W. Butzer, "The Realm of Nature and Society: Adaptation and Change in Historical Perspective," in B. L. Turner II et al. (eds.), *The Earth Transformed by Human Action*, New York: Cambridge University Press (Forthcoming), 1990.

2 I am indebted to Martin Parry for my appreciation of the importance of retrodiction.

3 I agree with Bennett that Whitehead's concepts of anticipation and the role of process in reality, the notion of culture, and the concept of adaptation essentially are interchangeable; see John Bennett, "Anticipation, Adaptation, and the Concept of Culture in Anthropology," *Science*, 192 (1976), 847-853; Alfred N. Whitehead, *Process and Reality*, New York: Macmillan, 1929.

Figure 1.1 Location Map of Equatorial Andes

subsistence and reproduction, and, in the process, furthering the creation of a meaningful world.

A processual approach to cultural ecology need not preclude interest in potentially stable adapted states. Indeed, this approach can be used to create scenarios of what such adapted states might look like, under given demographic, socio-economic, and environmental conditions. Such scenarios are generated in the present study, and compared against the evidence for highland Ecuador.

The Study Region

Our focus is on the northern and central highlands of Ecuador (Figures 1.1 and 1.2). Here can be found the archaeological remains of diverse agricultural fields and settlements, most of which date to before the Inca conquest. These highlands have remained the most densely populated part of the Ecuadorian state, with many subsistence or semi-subsistence farmers.

The northern and central highlands consist of high volcanic peaks and intervening valleys, mantled with volcanic ash. Most settlements are above 2,000 meters, as the lower slopes are frequently too steep, too dry or too moist for easy settlement. The upper limit of cultivation is about 3,800 meters; above this limit high grasslands (páramos) provide grazing.

The environment will be described in more detail in the following chapter. Wherever possible, environmental zones will be defined not on the basis of a priori criteria but rather in terms relevant to the major adaptive strategies: the "strategically relevant environment." In part, this will mean less emphasis on altitudinal zones, and more emphasis on soil slope, moisture, and fertility.

At present this region is occupied by three kinds of ethnic groups: indigenous Quichua groups, mestizo groups, and blacks. Although the three groups occupy distinct territories and share many agricultural practices, their access to resources and pursuit of non-agricultural options vary. Most of the information on traditional agricultural practices has been gathered from the Quichua groups.

Within the study region, ethnographic studies were undertaken primarily at Colta in Chimborazo province and at San Pablo, Atuntaqui, and Pimampiro in Imbabura province (Figure 1.2). Other locations served as sites for the study of agricultural productivity and prehistoric agricultural features. In this book, geographical scale will constantly shift as particular examples are used to help understand regional problems.

Traditional and Prehistoric Agriculture in the Andes

The study of past and present-day farming indicates the limitations and possibilities of the Andean environment and provides insights to prehistoric

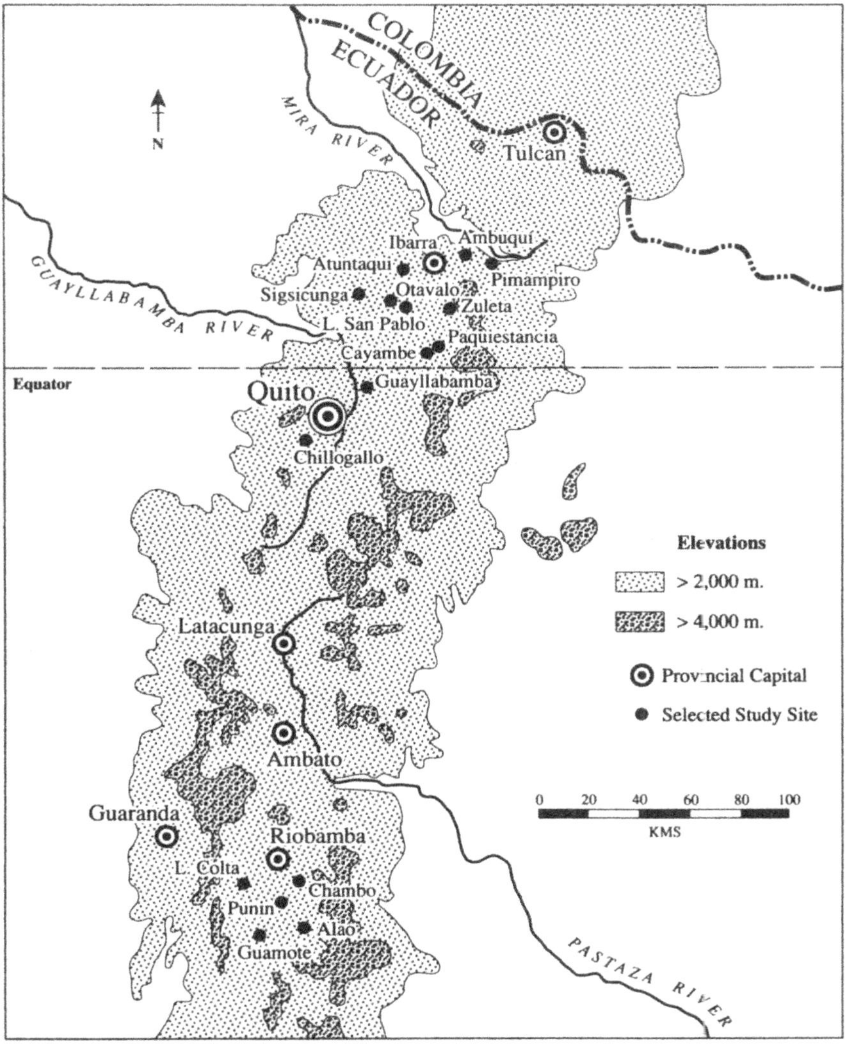

Figure 1.2 Location of Study Sites

farming patterns. Until recently, there have been few such studies. We often lack elementary details on the productivity and efficiency of subsistence agriculture, information vital to the prehistorian.[4] In the early 1970s, the field study of Andean agriculture was already under way, influenced by pioneering efforts of Carl Troll, Denevan at Wisconsin, Murra at Cornell, Mayer and Fonseca in Peru, and the Man and the Biosphere program of the United Nations, among others.[5] By the

---

4 Specific calls for research on adaptation in highland Ecuador may be found in John Stephen Athens, *Evolutionary Process in Complex Societies and the Late Period - Cara Occupation of Northern Highland Ecuador*, Ph.D. Dissertation, Ann Arbor: University Microfilms International, 1978, p.193; and in Leonard Field and Manuel Chiriboga, *Agricultura Andina*, Quito: CAAP, 1984.

5 See Carl Troll, "The Cordilleras of the Tropical Americas. Aspects of Climatic, Phytogeographical, and Agrarian Ecology," in Carl Troll (ed.), *Geo-Ecology of the Mountainous Regions of the Tropical Americas*, Bonn: Colloquium Geographicum Band 9, Ferd. Dümmlers Verlag, 1968, 13-56. Some of the earlier studies of Andean agricultural technology appear in Rogger Ravines (ed.), *Tecnología Andina*, Lima: Instituto de Estudios Peruanos, 1978, and in Heather Lechtman and Ana Maria Soldi (eds.), *La Tecnología en el Mundo Andino*, Mexico City: Universidad Nacional Autónoma de Mexico, 1981. See also the pioneering studies collected in César Fonseca and Enrique Mayer, *Comunidad y Producción*, Lima: FOMCIENCIAS, 1988. Another book-length study is Robin Donkin, *Agricultural Terracing in the Aboriginal New World*, Tucson: Viking Fund Publications in Anthropology, 1979. In the specific case of highland Ecuador, Roy Ryder had discovered raised fields at Cayambe in 1969 and at San Pablo in 1970 (with Alfred Siemens). Ryder also found abandoned semi-terraces near Cayambe, and David Preston found similar features near Pimampiro. An early report of these features is Roy Ryder, "El valor de la fotografía aérea en los estudios históricos y arqueológicos del Ecuador," *Revista Geográfica del IGM (Quito)*, 6 (1970), 40-42. Robin Donkin cites the relevant personal communication from Preston. An early study of prehispanic irrigation is the essay by Thomas Myers, "Evidence of Prehistoric Irrigation in Northern Ecuador," *Journal of Field Archaeology*, 1 (1974), 309-313. There are also references in colonial documents to Inca agricultural technology which will be cited later in this study. Valuable ethnographic information on traditional agriculture can be found scattered through a number of reports and articles, but modern field study of Andean agriculture in its cultural and geographical context was initiated by two graduates of the University of Wisconsin: Daniel Gade and Stephen Brush. See Daniel Gade, *Plants, Man and Land in the Vilcanota Valley of Peru*, The Hague: W. Junk, 1975, and Stephen Brush, *Mountain, Field and Family: the Economy and Human Ecology of an Andean Valley*, Philadelphia: University of Pennsylvania Press, 1977. A report on ongoing research by Steven Webster, "An Indigenous Quechua Community in Exploitation of Multiple Ecological Zones," *Actas y Memorias del XXXIX Congreso Internacional de Americanistas*, Lima, Vol. 3 (1971), 174-183, appeared about the same time. A different yet equally pioneering school of early research on Andean agriculture was published in Paul Baker and Michael Little (eds.), *Man in the Andes: A Multidisciplinary Study of the High-Altitude Quechua*, Stroudsburg, Pennsylvania: Dowden, Hutchinson, and Ross, 1976. For Ecuador, the pioneering work is David Basile, *Tillers of the Andes: Farmers and Farming in the Quito Basin*, Chapel Hill:

beginning of the decade of the 1980s there had been few detailed studies of Andean highland terracing, raised fields, or traditional irrigation, although all three technologies had already been shown to have been present in prehistoric Ecuador and Peru.

More recently, the study of prehistoric and traditional agriculture in the Andes has accelerated, building upon the earlier studies and including new approaches. Researchers from North America, Latin America, Europe, and Asia have kept in close contact, sharing ideas and information.

Although not all projects integrate archaeological, environmental, historical, and ethnographic components, it might be fair to say that such integration is increasingly expected. Noteworthy Peruvian projects, which have integrated many of these components, include several studies of terracing and irrigation in the Colca Valley by Denevan, Treacy, Guillet, and others;[6] work by Erickson, Kolata, Ortloff, and Flores Ochoa in the Lake Titicaca basin of southern Peru; by Zimmerer in the Paucartambo Valley;

University of North Carolina Department of Geography Studies in Geography 8, 1974, although much of interest may be found in the monographs in the *Llajta* series published in Quito.

6 See William Denevan (ed.), *The Cultural Ecology, Archaeology, and History of Terracing and Terrace Abandonment in the Colca Valley of Southern Peru*, Madison: Technical Report to the National Science Foundation and National Geographic Society (photocopy), 1986, and John Treacy, *The Fields of Coporaque: Agricultural Terracing and Water Management in the Colca Valley, Arequipa, Peru*, Ph.D. Dissertation,Ann Arbor: University Microfilms International, 1989. See also David Guillet, "Terracing and Irrigation in the Peruvian Highlands," *Current Anthropology*, 28 (1987), 409-430, Karl Zimmerer, *Seeds of Peasant Subsistence: Agrarian Structure, Crop Ecology and Quechua Agriculture in Reference to the Loss of Biological Diversity in the Southern Peruvian Andes*, Ph.D. Dissertation, Ann Arbor: University Microfilms International, 1988.

and by Sherbondy in Cuzco.[7] Reports on these projects, and other related studies, have been published in Peruvian journals and books.[8]

In Ecuador, similar work has been concentrated in the northern highlands, with many notable contributions.[9]

To simplify matters, we can view these studies of traditional and prehistoric Andean farming as addressing three challenges: (1) to develop a typology of Andean farming practices and relate these practices to ecological

---

7 See for example the articles published in William M. Denevan, Kent Mathewson and Gregory Knapp (eds.), *Pre-Hispanic Agricultural Fields in the Andean Region*, Oxford: BAR International Series 359, Two volumes, 1987; Jeanette Sherbondy, *The Canal Systems of Hanan Cuzco*, Ph.D. Dissertation, Ann Arbor: University Microfilms International, 1982; Zimmerer, *Seeds*; Alan Kolata, "The Agricultural Foundations of the Tiwanaku State: A View from the Heartland," *American Antiquity*, 51 (1986), 748-762; Alan Kolata, *La Tecnología y Organización de la Producción Agrícola en el Estado de Tiwanaku: Primer Informe de Resultados del Proyecto Wilajawira*, La Paz: Editorial Sui Generis, 1989; Clark Erickson, "Applications of Prehistoric Andean Technology: Experiments in Raised Field Agriculture, Huatta, Lake Titicaca, 1982-1983," in Ian Farrington (ed.), *Prehistoric Intensive Agriculture in the Tropics*, Oxford: B.A.R. International Series 232, 1985, 209-232; Clark Erickson, *An Archaeological Investigation of Raised Field Agriculture in the Lake Titicaca Basin of Peru*, Ph.D. Dissertation, Ann Arbor: University Microfilms International, 1988.

8 Especially noteworthy are Carlos de la Torre and Manuel Burga (eds.), *Andenes y Camellones en el Peru Andino*, Lima: Consejo Nacional de Ciencia y Tecnología, 1986, and Jeannette Sherbondy (ed.), *Antiguedad y Actualidad del Riego en los Andes*, Cusco: Instituto de Pastoral Andina, Allpanchis Numbers 26 and 27, 1986. A pivotal book with a purely contemporary focus is Enrique Mayer, *Land Use in the Andes: Ecology and Agriculture in the Mantaro Valley of Peru with Special Reference to Potatoes*, Lima: Centro Internacional de la Papa, 1979.

9 See especially Pierre Gondard and Freddy López, *Inventario Arqueológico Preliminar de los Andes Septentrionales del Ecuador*, Quito: MAG-PRONAREG-ORSTOM-Museo del Banco Central, 1983, Gregory Knapp, *Soil, Slope, and Water in the Equatorial Andes: A Study of Prehistoric Agricultural Adaptation*, Ph.D. Dissertation, Ann Arbor: University Microfilms International, 1984, and Patricia Mothes, *Pimampiro's Canal: Adaptation and Infrastructure in Northern Ecuador*, Austin: Unpublished M.A. thesis, Department of Geography, University of Texas at Austin, 1986. Of many articles by Chantal Caillavet, "Toponimia Histórica, Arqueología, y Formas Prehispánicas de Agricultura en la Región de Otavalo - Ecuador," *Boletín del Instituto Francés de Estudios Andinos (Lima)*, 12 (1983), 1-21, is especially pertinent. Also see Gregory Knapp and David Preston, "Evidence of Prehistoric Ditched Fields on Sloping Land in Northern Highland Ecuador," in William Denevan, Kent Matthewson and Gregory Knapp (eds.), *The Ecology and Archaeology of Prehispanic Agricultural Fields in the Central Andes*, Oxford: BAR International Series 359, 1987, 403-423. Additional materials of interest can be found in Ian Farrington (ed.), *Prehistoric Intensive Agriculture in the Tropics*, Oxford: BAR International Series 232, 1985, and in Janet Darch (ed.), *Drained Field Agriculture in Central and South America*, Oxford: BAR International Series 189, 1983.

niches; (2) to study the way Andean space is and has been organized; and (3) to explain the rationality of observed Andean patterns of land use and spatial organization, past and present.

Typologies of Land Use

There have been many attempts to define and name the significant environmental zones of the Andes and relate them to crops and cultural adaptations.[10] Special attention has been paid to John Murra's two systems of agriculture: potato farming in high elevations and maize farming in lower temperate elevations. According to Murra, the high temperate elevations in the central Andes are the zone of plants "domesticated locally, laboriously adapted to alpine conditions, grown on fallow land, and dependent on rainfall." The lower temperate elevations are associated with maize, grown on slopes, usually on irrigated, unfallowed terraces.[11] Numerous studies have focused on fallowed and/or rotated tuber production at higher elevations, and continuous maize production at lower elevations. Other crops, such as coca, may also be studied in terms

---

10 Indeed, the science of biogeography had its beginnings in the attempt to understand Andean altitudinal zonation. See Alexander Humboldt and Aimé Bonpland, *Essai sur la géographie des plantes; accompagné d'un tableau physique des régions equinoxiales*, Mexico City: Institut Panamericain de Géographie et d'Histoire, Editorial Cultura, 1955. The eight natural regions of Peru defined by Pulgar Vidal were a pioneering attempt at ethnoscientific exploration of the totality of the Andes, not just the highlands: Javier Pulgar Vidal, *Historia y geografía del Perú: Tomo I, Las ocho regiones naturales del Perú*, Lima: Universidad Nacional Mayor de San Marcos, 1946. Carl Troll introduced a true geoecological approach with his three-dimensional assessment of environmental zones in the Andes as a function of latitude as well as altitude: Troll, "Cordilleras." Another system based on a presumed regular relationship between temperature, precipitation, latitude, and altitude on vegetation is that of Leslie R. Holdridge, *Life Zone Ecology*, San José, Costa Rica: Tropical Science Center, 1967. This system was first applied to the Andes by Joseph A. Tosi Jr., *Zonas de Vida Natural en el Perú*, Lima: Instituto Interamericano de Ciencias Agrícolas de la OEA: Zona Andina, 1960. It has since been applied to the other Andean countries, including Ecuador. The first attempt to synthesize this literature for several countries is James Parsons, "The Northern Andean Environment," *Mountain Research and Development*, 2 (1982), 253-262. For a French geographer's more recent synthesis of Troll and Pulgar Vidal, see Olivier Dollfus, *El Reto del Espacio Andino*, Lima: Instituto de Estudios Peruanos, 1981.

11 This distinction is developed in John Murra, "Rite and Crop in the Inca State," in S. Diamond (ed.), *Culture and History: Essays in Honor of Paul Radin*, New York: Columbia University Press, 1960, 393-407. The quotation is found on page 401. The spiritual ancestor of Murra's discussion is the remarkable essay by Carl Sauer, "Cultivated Plants of South and Central America," in Julian Steward (ed.), *Handbook of South American Indians*, Washington: Smithsonian Institution Bureau of American Ethnology Bulletin 143, 1950, 487-543.

of altitudinal zonation.[12] These approaches do not however provide for a complete understanding of Andean agriculture. The focus on crops (rather than calorie production and labor costs), and the neglect of alternative uses of given niches, makes it difficult to interpret observed patterns as a result of historically variable processes. Some readers could conclude from these studies that what exists today is optimal and is determined by the environment.

In recent years, taxonomies of Andean land use have increasingly taken into account the notably conspicuous artificial agricultural landforms. These may include terraces, raised fields, canals, and other features. Many agricultural landforms are abandoned. Possibilities of re-utilization cannot be assessed, however, without studying the adaptive context of the technologies in question.[13]

Studies of Spatial Organization

It has been argued that Andean farmers have attempted to gain access to all altitudinal zones, from the subtropical to the high pastures above the limit of farming. This objective is called "verticality" or, more recently, ecological complementarity.[14] It is sometimes assumed that Andean farmers do not welcome land, labor, or commodity markets but rather have preferred institutions of reciprocity or redistribution to gain access to culturally important crops. It has been further argued that Andean settlements are most efficiently located on the boundaries between natural ecological zones, and archipelagos of ethnic islands can be expected to

12 See, for example, Daniel Gade, *Daniel Gade, Plant Use and Folk Agriculture in the Vilcanota Valley of Peru: A Cultural-Historical Geography of Plant Resources*, Ph.D. Dissertation, Ann Arbor: University Microfilms International, 1967, especially page 167; Stephen Brush, *Subsistence Strategies and Vertical Ecology in an Andean Community: Ushucmarca, Peru*, Ph.D. Dissertation, Ann Arbor: University Microfilms International, 1973, page 155; Mayer, *Land Use*, pages 40-41, and Basile, *Tillers*.

13 A point made forcefully by César Fonseca Martel, "Destruccíon de Andenes de la Cuenca del Río Cañete," in Carlos de la Torre and Manuel Burga (eds.), *Andenes y Camellones en el Perú Andino*, Lima: Consejo Nacional de Ciencia y Tecnología, 1986, 361-368.

14 See Webster, "Indigenous Quechua Community," and the classic formulation of the concept by John Murra, "El 'Control Vertical' de un Máximo de Pisos Ecológicos en la Economía de las Sociedades Andinas," in John Murra (ed.), *Visita de la Provincia de León de Huánuco en 1562*, Huánuco, Peru: Universidad Nacional Hermilio Valdizán, 1972, 429-476. The concept of ecological complementarity, meant to generalize the concept beyond altitudinal zones, is developed in Shozo Masuda, Izumi Shimada and Craig Morris (eds.), *Andean Ecology and Civilization*, Tokyo: University of Tokyo Press, 1985; see especially the introductory essay by Shimada.

emerge where key resources are concentrated in remote areas. Thanks to the studies of verticality, considerable contributions to our knowledge of the spatial organization of settlements, ethnic units, and economic links in the Andean world, past and present, have been made. As the research has progressed, it has become apparent that markets have played a part in Andean adaptation, even in pre-Hispanic times, especially in areas away from the heartland of the Inca empire in southern Peru. The alternatives for defining and controlling ecological zones have been very diverse indeed.[15]

Andean Rationality

A deficiency of research on the Andes has been the lack of convincing explanations of Andean behavior. The verticality tradition tends to assume an inherent cultural habit of seeking a maximum number of ecological zones, but there is no obvious reason why farmers should strive to maximize niche access, as opposed to resource access in general or optimization in general.[16]

While verticality can provide a variety of consumption goods, variety can also be obtained through local agricultural diversification, which has the further advantage of minimizing transport costs, limiting pest proliferation, and optimizing soil nutrient use.

The use of multiple niches has been justified as eliminating seasonal unemployment, in effect invoking "time geography" as an explanatory approach in understanding Andean land use. There is as yet little evidence for this. Furthermore, it is at least unclear whether farmers prefer continuous work to seasonal "unemployment" providing opportunities for off-farm activities. These matters are discussed further in Chapter 4.[17]

Another criterion put forward to explain Andean rationality has been energy efficiency. It has been argued that "groups with the highest long-term energy efficiency may be regarded as the most adaptive," where energy efficiency is defined as the ratio between energy production and energy expenditure. Thomas has suggested that this approach helps explain why Andean farmers sit down whenever possible, employ children for herding

---

15 See Stephen Brush, "Man's Use of Andean Ecosystem," *Human Ecology*, 4 (1976), 147-166, and Frank Salomon, *Native Lords of Quito in the Age of the Incas: The Political Economy of North Andean Chiefdoms*, Cambridge: Cambridge University Press, 1986, for discussions of the varieties of archipelagos, marketing centers, and trading arrangements in the Andes.

16 A point made by Rodrigo Sánchez, "The Model of Verticality in the Andean Economy: A Critical Reconsideration," *Actes du XLIIe Congrès International Des Americanistes*, Paris, 1976, Vol. IV, 213-231.

17 See Jürgen Golte, *La Racionalidad de la Organización Andina*, Lima: Instituto de Estudios Peruanos, 1987; and, for a general outline of time geography and its significance, see Tommy Carlstein, *Time Resources, Society, and Ecology*, London: George Allen & Unwin, 1982.

tasks, and use animals rather than humans for carriage.[18] These particular examples however can better be seen as results of the "principle of least effort": in the short run, where land, capital, and other resources are fixed and goals have been set, one tends to minimize labor time (especially unpleasant labor time) in meeting one's goals. This principle is so often valid because there are alternative uses to labor time: labor saved in production can be used in enhancing consumption, competition, and reproduction, insofar as these are separable from production.[19] In contrast, there is little evidence to support the notion that societies evolve towards states of greater overall energetic efficiency (the reverse could in fact be argued). Concepts of efficiency may help to explain the means people employ to reach their goals in the political and cultural arenas, but do little to explain the details of the goals themselves.

To a certain extent goals, in the Andes and elsewhere, are a product of population density, itself a product of the often unpredictable vagaries of demographic history. But goals are also linked to the ways groups cohere, compete, and reproduce, and these political and cultural factors are also best studied in terms of their own historical trajectories. The salience of history suggests that the interaction of goals and means can perhaps best be studied in evolutionary terms rather than in systems frameworks, valuable as the latter may be as heuristic devices.

Despite the use of modelling, the present book is less a study of "cultural ecosystems" than a study of adaptive dynamics: the actual diversity of options facing Andean decision-makers in coping with a diverse environment, and the characteristics of these options which make some more attractive than others under specific environmental, demographic and socio-economic contexts.

---

18 See R. Brooke Thomas, "Energy Flow at High Altitude," in Paul Baker and Michael Little (ed.), *Man in the Andes: A Multidisciplinary Study of the High-Altitude Quechua*, Stroudsburg, Pennsylvania: Dowden, Hutchinson, and Ross, 1976, 382.

19 See George Kingsley Zipf, *Human Behavior and the Principle of Least Effort: An Introduction to Human Ecology*, Cambridge: Addison-Wesley Press, 1949. Without subscribing to Zipf's ideas in detail, it remains true that a large proportion of explanation in social science is based on an assumption of effort minimizing behavior in reaching goals (not in defining goals). For an excellent more recent discussion of efficiency in cultural ecology, see Michael A. Jochim, *Strategies for Survival: Cultural Behavior in an Ecological Context*, New York: Academic Press, 1981. Although a culture where production is indistinguishable from reproduction, consumption, or social interaction is theoretically possible, here it is assumed that in all known societies the means-end distinction has had some (although not absolute) validity, albeit as a continuum.

Adaptive Dynamics

For many, there is a special fascination in studying those culture traits most closely linked to the survival and reproduction of human groups. Such traits and related behaviors have long been the objects of study for cultural ecology.[20]

In the 1960s and early 1970s, cultural ecology attained maturity as the study of subsistence producers in the light of evolutionary and especially ecological systems theory.[21] In retrospect, the pioneering studies in cultural ecology were firmly grounded in field work, were conscious of the role of large scale economic context in setting some major parameters for adaptation, and did not study societies in a timeless present. Nevertheless, in the late 1960s and early 1970s, cultural ecology came to be increasingly influenced by general systems theory and systems ecology.[22] As developed by some, cultural ecology became the art of translating human facts into ecological systems terminology. The human economy was conceptualized in terms of energy flows, with special attention to feedback loops assuring stability. Practical research in "cultural ecosystemicism" (to use Bennett's phrase) tended to be oriented toward proving that apparently irrational or ceremonial behavior actually served to maintain cultural ecosystems.[23]

Probably few were ever true "ecosystemicists;" rather, ecosytemicism was a recurrent characteristic of the literature of the time, often adhered to without thinking through its implications, and often coexisting in the same text with other, inconsistent, perspectives. Ecosystemicism can be a sort of neo-functionalism with attendant problems of logical circularity. The peculiar characteristics of human ecosystems (open, with poorly defined or constantly changing boundaries, subject to constant novelty through external catastrophic input or changing choices of the participating individual human actors) often subvert efforts to find homeostatic mechanisms. Thus the concept of human ecosystem is probably best seen as a heuristic device, an aid to conceptualizing the range of mutual

---

20 See Julian Steward, "Ecological Aspects of Southwestern Society," *Anthropos*, 32 (1937), 87-104, and Julian Steward, *Theory of Culture Change: The Methodology of Multilinear Evolution*, Urbana: University of Illinois Press, 1955.

21 Clifford Geertz, *Agricultural Involution: The Process of Ecological Change in Indonesia*, Berkeley: University of California Press, 1963; H.C. Brookfield and Paula Brown, *Struggle for Land: Agriculture and Group Territories among the Chimbu of the New Guinea Highlands*, Cambridge: Cambridge University Press, 1963.

22 Eugene P. Odum, *Fundamentals of Ecology*, Philadelphia: Saunders, 1953.

23 See for example Roy Rappaport, *Pigs for the Ancestors: Ritual in the Ecology of a New Guinea People*, New Haven: Yale University Press, 1967; Marvin Harris, *Cows, Pigs, Wars and Witches: the Riddles of Culture*, New York: Random House, 1974. The term "cultural ecosystemicism" is found in John Bennett, *The Ecological Transition: Cultural Anthropology and Human Adaptation*, New York: Pergamon, 1976, page 164.

interactions between people, plants, animals, and other components of the world.[24] This heuristic usage appears to constitute one of the major tasks of cultural ecology in the 1990s, but there remain two additional problems: (1) the term system has an inherent mechanical, timeless and spaceless connotation which may cause some to prefer older (if less scientifically negotiable) terms for interactive complexes such as region, cultural landscape, place, or world;[25] and (2) the systems framework should not be allowed to discourage analytic research on the components of adaptation and the micro-scale processes of choice between alternatives which may indeed contribute explanations for observed cultural ecological patterns.

If cybernetics does not supply the only appropriate framework for understanding human-environment relations, there is no lack of alternatives. One approach has been to look to evolutionary theory and biology for analytic frameworks. This has meant going beyond "adaptational rhetoric" to define the fundamental principles of the adaptation process. Here, a distinction should be made between adaptation and adaptedness.[26] Cultural adaptation is the modification of culture, in a particular adaptive context; cultural adaptedness is a momentarily stable relationship between culture and environment. Complete adaptedness may be rare; individuals and groups are constantly selecting and modifying behaviors as prior decisions or external events change the social and physical environment, as new options become available, and as information and meaning are redefined in thought, art, and speech.

Cultural evolution should be carefully distinguished from human evolution. In cultural evolution, selective pressures[27] operate in the first instance on components of culture (learned behaviors, associated tools, ideas). Selection occurs both on the cognitive and behavioral levels,[28] so that adaptation can occur much more rapidly than biological adaptation. At

---

24 See Benjamin Orlove, "Ecological Anthropology," *Annual Review of Anthropology*, 9 (1980), 235-273; the useful essays in Emilio Moran (ed.), *The Ecosystem Concept in Anthropology*, ed. Emilio Moran, Boulder: Westview Press, 1984; Patrick Kirch, "The Archaeological Study of Adaptation: Theoretical and Methodological Issues," in Michael Schiffer (ed.), *Advances in Archaeological Method and Theory*, New York: Academic Press, 1980, 101-156; Michael Jochim, "Breaking Down the System: Recent Ecological Approaches in Archaeology," *Advances in Archaeological Method and Theory*, 2 (1979), 77-116. See too B.L. Turner and Stephen Brush (ed.), *Comparative Farming Systems*, New York: The Guilford Press, 1987.

25 It is interesting to observe, simply from a perspective of terminology, the emergence of "landscape ecology."

26 Patrick Kirch, "Archaeological Study," 101. For an example of the confusion the term can still engender, see Thomas Bargatzky, "Culture, Environment, and the Ills of Adaptationism," *Current Anthropology*, 25 (1984), 399-406, and the comments in the same issue.

27 Kirch, "Archaeological Study," 104, 113, 115.

28 A. Alland Jr.,"Adaptation," *Annual Review of Anthropology*, 4 (1975), 59-73.

the same time, numerous apparently contradictory or inefficient cultural traits may survive for a variety of reasons, including inertia (as habits), as secondary techniques or as options for unique situations.[29]

The biological reproductive success of the culture-bearing population is therefore less relevant to cultural survival than it is for the survival of the population's genes. However, contrary to Sahlins, adaptation is not merely a stipulation of "minimum positive functioning," where "anything that does not subject the people or the system to material destruction, is advantageous from the point of view of adaptation."[30] Every set of strategies which assure minimum positive functioning has a labor and capital cost, a limit of flexibility, and a risk of failure, so that it is in people's interest to constantly compare and assess alternative subsistence options. Although means are assessed in terms of goals, goals may also be assessed in part on the basis of experience derived from their realization. Adaptation is not a simple initial process which, once accomplished, ceases, to allow the real business of culture to proceed, but rather involves continual choice, continual competitive evaluation of alternative cultural traits with respect to efficiency, flexibility, and coevolution potential; it can in the broadest sense be seen as a synonym for culture itself.[31]

Choice depends on the variety of alternatives available. Variation is favored by invention and diffusion, the classical explanations used by geographers to explain change; however, the actually observed level of variation depends not only on the rate of introduction of new traits, but also on the rate of extinction of old traits. It seems plausible that a diverse and uncertain natural and social environment favors survival of diverse traits,[32] so that mountain cultures such as those to be found in the Andes should be ideal for the study of adaptive dynamics.[33] Historical factors of contact and conflict between diverse groups seem often to result in a sorting out of distinct ethnic groups with distinct adaptive strategies in distinctive niches, increasing the local pool of adaptive alternatives.[34] Choice is thus normally possible and meaningful.

From the perspective of short term adaptive dynamics, as problems and opportunities emerge people find solutions by choosing among a number of

---

29 William Denevan, "Adaptation, Variation, and Cultural Geography," *Professional Geographer*, 35 (1983), 402.

30 Marshall Sahlins, *Culture and Practical Reason*, Chicago: University of Chicago Press, 1976, 209.

31 Bennet, "Anticipation."

32 Kirch, "Archaeological Study," 116.

33 See the studies assembled in Nigel Allan, Gregory Knapp and Christoph Stadel (eds.), *Human Impact on Mountains*, Totowa, New Jersey: Rowman and Littlefield, 1988.

34 Fredrik Barth, "Ecologic Relationships of Ethnic Groups in Swat, North Pakistan," *American Anthropologist*, 58 (1956), 1079-1089. John W. Bennett, *Northern Plainsmen: Adaptive Strategy and Agrarian Life*, Chicago: Aldine, 1969.

available alternative strategies, basing their decision in part on the relative efficiencies of the alternatives. Their choices in turn may alter the social and natural environment and may lead to further problems or opportunities. The decision-making units and institutions, and the salient criteria of efficiency, are variable;[35] but within a given socio-economic context, "natural and social environments may be examined as sources of problems and the means to their solution. Budgets ... may be constructed in order to assess options...."[36] For example, a subsistence farmer's choice of crop may in part be conditioned by the competing labor efficiencies of the alternatives in providing a family with culturally acceptable calories. Models of crop yields and labor inputs under different management regimes, climate and soil characteristics, and crop characteristics can be used to help understand actual choices of adaptive strategy and actual variations in land use across landscapes.

There is no single definition of the relevant environment for human adaptation, since different adaptive strategies are sensitive to different components of the physical context, and the significant strategies are conditioned by the social context. Nevertheless, some studies have defined biomes, biotopes, or ecozones in an intuitive or ad hoc fashion prior to proceeding with a description of human behavior.[37] Soil classification systems are perhaps most relevant to agriculture, but the common classification schemes, such as the USDA system, are not designed to be predictive of yield or crop behavior.[38]

An alternative is an "adjoint methodology" in which the strategically relevant environment is defined in tandem with (socially and culturally conditioned) potential crops and cropping methods.[39] Models of adaptive

---

35 For attempts to grapple with issues of the interplay between social structure, ecology, and adaptation see, for example, Michael Watts, *Silent Violence: Food, Famine and Peasantry in Northern Nigeria*, Berkeley: University of California Press, 1983, Lawrence Grossman, *Peasants, Subsistence Ecology, and Development in the Highlands of Papua New Guinea*, Princeton: Princeton University Press, 1984, P. Blakie and H.C. Brookfield (eds.), *Land Degradation and Society*, London: Methuen, 1987, and Zimmerer, *Seeds*.

36 Jochim, *Strategies*, 104.

37 Bruce Winterhalder, "Environmental Analysis in Human Evolution and Adaptation Research," *Human Ecology*, 8 (1980), 135-170.

38 USDA (United States Department of Agriculture), *Soil Taxonomy*, Washington: U.S. Government Printing Office, 1975; Pedro Sanchez, Walter Couto and Stanley Buol, "The Fertility Capability Soil Classification System: Interpretation, Applicability, and Modification," *Geoderma*, 27 (1982), 283-309.

39 Martin Parry and Tim Carter have defined the term "adjoint methodology" to refer to methods which define environmental problems by beginning with society (as opposed to the "direct" methodologies of the natural scientist who begins with the environment and then looks at the effects of environmental perturbations: Martin Parry and Tim Carter, *Assessing the Impact of Climate Change in Cold Regions*, Laxenburg: International Institute for Applied Systems Analysis, 1984.

strategies and strategically relevant environments can be calibrated, tested and used in a variety of ways. Interviews with field informants and actual field experiments can help determine yield, efficiency, and limitations of agricultural practices. Efficient solutions to the problems facing historic or prehistoric populations can be "retrodicted" and then checked against what is known of early land use, through early chronicles, environmental archaeology, the location of abandoned agricultural features, and house sites. Appropriate technical solutions to current problems can be suggested and the results monitored. Thus the further development of adaptive dynamics and cultural ecology is intimately linked with the further development of studies of agricultural history and prehistoric agriculture, as well as with the building of ecologically and culturally appropriate development strategies.

This book will characterize the changing adaptive strategies of highland Ecuador, determining the strategically relevant environments. Models of strategies will be checked against the contemporary situation and especially by comparing a hypothetical reconstruction of pre-Hispanic agricultural patterns in the Ecuadorian highlands against data on prehistoric settlement and land use. It will be shown that some measure of prediction and retrodiction is achieved, indicating that some of the processes of culture and adaptation in the Andes have been clarified.

Chapter 2

The Strategically Relevant Physical Environment

The relevant physical environment for contemporary Andean farmers includes those characteristics of Andean climate, soils, and organisms which can affect average crop yields and their variance. The more that is known about the effects of varying environmental characteristics on yields, the greater the potential ability to understand the basis for observed farmer choice of adaptive strategy.

Each crop variety and agricultural strategy has a production function which specifies the relationship between environmental parameters, labor, capital, technical practices and yield. If there were only one crop and strategy, production would be uniquely determined by environmental factors, and could be readily predicted on the basis of a simple map of the relevant environment and a knowledge of the degree to which the farming population had fully occupied its niche.

There are however multiple crops and tactics which Andean farmers may employ, as will become clearer later in this book. Their production functions differ not only in elasticity with respect to environmental parameters, but also with respect to labor and capital inputs. The availability of options with differing production functions can result in a continual process of adaptation, where more inputs are devoted to the best land and the type of land considered best depends in part on the global availability of inputs and level of demand.[1]

This chapter will examine the productivity of some of the basic crop-growing strategies of the Andes and in the process define strategically relevant environments, using an adjoint methodology. The next chapter will then investigate the range of tactical adaptability to variations in demand and inputs.

---

1 Most inputs should be devoted to the most productive land simply because the marginal rate of return should initially be highest there. If the marginal rate of return to an input is not constant over space, the redistribution of inputs should provide a "free" global increase in production. This of course does not determine that such a distribution will in fact take place, but the failure to do so requires an explanation.

An Initial Look at Climate and Soil

It is not possible to separate a description of the environment from the social and strategic context in which it is relevant. Before analyzing in detail the parts of the environment of interest to local farmers, it may be helpful to provide a general overview of the area's climate and soils from an external, scientific perspective, postponing a more detailed discussion of adaptive strategies until the next section.

Precipitation, temperature, and soil vary continuously in the northern and central highlands of Ecuador, as a function of varying altitude, slope, and topographic position; soils vary also as a function of parent material, vegetation, and history. These highlands, which for brevity we will call the equatorial Andes, form a continuous high chain dividing the Amazon lowlands in the east from the lowland coastal valleys in the west. The coastal valleys and Amazon basin are under 300 meters elevation; the Andes rise to a maximum of 6,310 meters (at Chimborazo). At no location is it possible to cross from the Amazon to the Pacific without crossing a pass over 3,000 meters high, and there is only one place where it is possible to do so under 3,500 meters.[2]

The highlands consist of two north-south trending volcanic mountain chains (*cordilleras*) separated by inter-Andean valleys. The cordilleras are only breached by the three rivers draining the inter-Andean valleys: the Mira, the Guayllabamba, and the Pastaza. These rivers define three inter-Andean basins separated by transverse uplands (*nudos*) at the interfluves. Typically, the inter-Andean basins are flat to gently sloping at their centers, filled with alluvial, volcanic, and glacial deposits, while the outer flanks of the highlands are highly dissected and offer little flat land. In a few cases, however, faulting has created small north-south trending valleys on the western flank of the highlands which approximate the inter-Andean valleys in offering gently sloping lands at high elevations.[3]

PRECIPITATION

Average annual rainfall in the northern and central highlands varies from under 300 mm to over 3,000 mm, although values of over 1,500 mm are generally limited to the eastern and western flanks of the highlands.

The quantity of rain varies with aspect and topographic position; as a general rule, the driest areas are the centers of the inter-Andean basins and the canyons where the rivers trench through the flanking cordilleras. The wettest areas are the lower western slope of the western cordillera and the lower eastern slope of the eastern cordillera (the "outer flanks"). This is because for at least part of the year, both sides of the highlands are under the influence of unstable moist tropical air masses. The inter-Andean

---

[2] At Alausí.

[3] At Guaranda, Alausí, and Sigchos.

valleys are partially insulated from the effects of these air masses, and experience local convective thundershowers driven by local air circulation with uplift on the valley margins and stagnant descending air in the valley centers.[4]

Because of the diverse processes involved, there is no single rule relating precipitation to elevation. On the outer flanks, precipitation tends to decrease with increasing elevation, while in the intermontane valleys distance from valley center is a better predictor of rainfall than is elevation. Average precipitation may change rapidly over short distances, providing special challenges for agricultural adaptation. This is especially the case where average precipitation dips below the critical minimum value for successful unirrigated agriculture.

Precipitation in the highlands varies seasonally in accordance with the changing factors encouraging atmospheric instability. In general, the Amazon basin portion of Ecuador is always under the influence of unstable moist tropical air masses and has no pronounced dry season. The western coastal area, however, is influenced by dry air masses from the south and cool offshore currents from June to December, the coastal dry season.

Seasonal patterns of highland rainfall differ from those in adjacent lowlands and are probably correlated with seasonal variations in wind directions and strengths, and with variations in insolation due to varying solar altitude.[5] Data for most highland stations show a double-peaked pattern of seasonal rainfall, with the peaks corresponding to the equinoxes. The dry season associated with the June solstice (summer, *verano*) is more intense and prolonged than that associated with the December solstice (little summer, *veranillo*).[6]

---

4 See A. M. Johnson, "The Climate of Peru, Bolivia, and Ecuador," in W. Schwerdtfeger (ed.), *Climates of Central and South America*, Amsterdam: Elsevier, 1976, 147–218; M. Frère, J. Rea and J. Q. Rijks, *Estudio Agroclimatológico de la Zona Andina*, Rome: Food and Agriculture Organization, 1975; Pierre Pourrut, "Los Climas del Ecuador: Fundamentos Explicativos," in *Los Climas del Ecuador*, Quito: Centro Ecuatoriano de Investigación Geográfica, Documentos de Investigación No. 4, 1983, 7–41.

5 See Edwin Ferdon, Malcolm Bissell and William Steere, *Studies in Ecuadorian Geography*, Santa Fe, New Mexico: School of American Research Monograph Number 15, 1950. Near the equator, the sun is directly overhead at noon about the equinoxes, presumably encouraging precipitation peaks at these periods. The importance of local convective storms at Quito is indicated by the timing of precipitation; the highest percentage of rainfall occurs from noon to 4 p.m. when temperatures are highest. Occasionally precipitation is affected by squall lines associated with cold fronts penetrating from the south, especially during the September–December period (Diego del Alcazar, personal communication).

6 This difference may be due to the influence of the high pressure cell and dry air masses to the south and west after the June solstice.

TEMPERATURE

Mean annual temperature decreases with increasing altitude by about 5.7 degrees Celsius per 1,000 meters.[7] Temperatures are about one degree Celsius higher in dry areas.[8] These lapse rates are calculated with data from stations on flats or gentle slopes. There is little information for mean annual temperature on steep slopes, so that the effects of aspect are not known; the equatorial latitude presumably implies little relationship between aspect and annual mean temperatures.

As is generally true in the tropics, seasonal differences in temperature are less than diurnal differences. There is normally less than a 1.5 degree difference between the coldest and warmest months.[9] The daily cycle of temperatures, however, is normally on the order of 10 degrees or greater.[10] Frosts are a special problem. Frosts are almost entirely nocturnal, due to net radiation losses coupled with cold air drainage (*heladas blancas*), since cold fronts with dramatic temperature effects (*heladas negras*) normally do not penetrate this far north. The diminished cloud cover in the dryer seasons facilitates increased frost frequency. Thus, frost frequency and frost severity peak in the June–September dry season, with a secondary peak around January. Probably most of the equatorial Andes above 2,000 meters elevation experiences some risk of frost. Frosts are normally not very profound, however; the lowest ground surface temperature recorded at Quito observatory from 1891 to 1975 was –4.9 degrees.[11]

---

7 Using mean data for highland stations in northern Ecuador for the 1964–1973 period, the normal lapse rate for temperature (T, in degrees Celsius) as a function of altitude (A, in km.) is $T = 28.4 - 5.7A$. See PRONAREG-ORSTOM (Programa Nacional de Regionalización Agraria - Office de la Recherche Scientifique et Technique Outre-Mer), *Estudio Hidro-Meteorológico e Hidrogeológico Preliminar de las Cuencas de los Ríos Esmeraldas y del Norte Ecuatoriano*, Quito, 1977. For the general Andean case, see Frère, Rea, and Rijks, *Estudio*. Cañadas found a slightly lower lapse rate (5 degrees) by including subtropical and tropical stations in his regression: Luis Cañadas, *El Mapa Bioclimático y Ecológico del Ecuador*, Quito: Banco Central del Ecuador, 1983.

8 That is, in areas with under 600 mm annual precipitation. See PRONAREG-ORSTOM, *Estudio*.

9 See PRONAREG-ORSTOM, *Estudio*, Volumen Anexo, 9.

10 At Izobamba, to the south of Quito, in 1981 the average daytime maximum was 11.6 degrees higher than the average daytime minimum. Daytime maxima varied less seasonally than daytime minima. See INAMHI (Instituto Nacional de Meteorología e Hidrología), *Anuario Meteorológico 1981*, Quito, 1983, 26–37.

11 November 17, 1948. See *Heladas Registradas en la Estación Meteorológica de Quito*, manuscript in the Quito Observatory meteorological station.

EVAPOTRANSPIRATION

The amount of water required by a crop can be estimated by calculating the potential evapotranspiration (PET), which is a function of temperature, insolation, wind speed, and humidity. Although some authors imply otherwise, evapotranspiration declines only slightly with increasing elevation in the equatorial Andes, but it does vary seasonally, reaching a maximum in the dry summer.[12]

SOIL

The equatorial Andes are especially appropriate for a study of this type because of a relatively uniform soil parent material. Most soils are derived from volcanic ash, subject in varying degrees to post-deposition weathering, erosion, and redeposition.

Recent studies by Ramón Vera and Chalmers Clapperton have contributed to an improved understanding of the characteristics of highland volcanic ash. Most of the underlying material belongs to the Cangahua Formation, made up of (1) ash falls without much wind transport, (2) mudflows (lahars), (3) soil horizons developed on the preceding materials, and (4) reworked materials derived by later erosion. Generally, the original ash is fine sandy in texture, but soil-forming processes have produced clays. The ash appears to have been laid down between 120,000 and 10,000 years ago.[13]

Soil development on the Cangahua Formation and subsequent layers of primary ash and reworked materials has typically resulted in the creation of black topsoils rich in organic matter. There has often been clay translocation into a subsurface horizon. Quite frequently one encounters layers which have been converted to hardpan (duripan), locally called

---

12 PET, calculated by the Penman method, decreased only 2 mm. for every 100 m. increase in elevation: Luis Cañadas and H. Salvador, *Agrometeorological Assessment Models for Rural Development in the Central Sierra of Ecuador*, Columbia, Missouri: Center for Environmental Assessment Services, 1982. This is even less than the lapse rate reported for the northern Andes by Frère, Rea, and Rijks, *Estudio*, 133. Analysis of Piché Evaporimeter data from 20 stations in the northern Ecuadorian highlands (1,730 to 3,860 meters) shows that the 10 stations above 2,800 meters average 1,061 mm/year PET; the 10 stations below 2,800 meters average 1,086 mm/year. For the data, see PRONAREG-ORSTOM, Estudio Hidro-Meteorológico, Annex, 17–18.

13 See Chalmers Clapperton and Ramón Vera, "La Secuencia Glacial del Cuaternario en el Ecuador, una Interpretación al Modelo de W. Sauer," *Paisajes Geográficos (Quito)*, 16 (1986), 3–20, and Ramón Vera, "El Origen de Cangahua," *Paisajes Geográficos (Quito)*, 16 (1986), 21–28 The classic account by Walther Sauer, *Geología del Ecuador*, Quito: Ministerio de Educación, 1965 is now partly obsolete.

cangahua.[14] Soils may vary then from prairie soils with thick black topsoils to outcrops of exposed duripan or even sand dunes, produced by wind and water erosion.[15]

TABLE 2.1 MEASURED SOIL FERTILITY DATA, VARIOUS SITES

			Nutrients (kg./ha.)					
STUDY	CROP	NUMBER OF SITES	PH		NITROGEN		PHOSPHATE (P2O5)	
			Mean	*C.V.*	*Mean*	*C.V.*	*Mean*	*C.V.*
Westra	Barley	70	6.6	.06	182	0.39	76	0.92
Bejarano	Potatoes	5	6.2	.06	90	0.19	37	0.23
Mothes	Unirrigated beans, corn	2	6.5	—	2.2	—	28	—
Mothes	Irrigated	5	7.6	.06	12	1.0	98	0.62
PIP	Maize	4	6.5	.03	72	0.19	97	0.26
San Jose type profile (flats)	Pasture	1	5.6	—	—	—	120	—
Knapp	Virgin steep slope	1	6.5	—	146	—	101	—
Knapp	Intensive maize, steep slope	2	6.4	—	96	—	21	—

14 An early report is that by Erwin Frei, *Informe al Gobierno del Ecuador sobre Reconocimientos Edafológicos Exploratorios*, Rome and Quito: Food and Agriculture Organization (Mimeographed), 1957. Most of our modern knowledge of Ecuadorian soils is due to the efforts of a team of French soil scientists led by F. Colmet Daage; see for example, F. Colmet Daage et al., *Caracteristiques de Quelques Sols D'Equateur Derivés de Cendres Volcaniques*, Quito: Programa Nacional de Regionalización Agrícola (Mimeographed), 1968. A selective report of the French team and their Ecuadorian collaborators is PRONAREG-ORSTOM (Programa Nacional de Regionalización Agraria - Office de la Recherche Scientifique et Technique Outre-Mer), *Mapas de Suelos por Regionalización*, Martinique and Quito: Centre des Antilles - Bureau des Sols, 1979.

15 In technical language, the soils are generally Mollisols or Eutrandepts, with some Dystrandepts at high elevations. Moisture regimes are udic or ustic. Studies of soil erosion in Ecuador may be found in CEDIG (Centro Ecuatoriano de Investigación Geográfica), *La Erosión en el Ecuador, Documentos de Investigación No. 6*, Quito, 1986. Also see the map of erosive processes, PRONAREG-ORSTOM (Programa Nacional de Regionalización Agraria - Office de la Recherche Scientifique et Technique Outre-Mer), *Los Principales Procesos Erosivos en Ecuador*, Quito, 1984.

Table 2.1 indicates the measured soil fertility characteristics of a number of sites in the equatorial Andes.[16] In the qualitative interpretations of the soil laboratory, nitrogen levels were generally medium, with relatively little variation. The most likely limiting factor is phosphorus.[17]

There is evidence that soil phosphorus is higher in flats than on slopes; values are medium rather than low.[18] Gentle slopes and flats are probably the sites of nutrient accumulation, washed in from adjacent slopes.

Heavily cropped unirrigated fields on slopes have the lowest phosphate levels, as in the value of 21 kilograms/hectare observed on frequently cropped slopes near La Vega. The Bejarano sites provide a view of intermediate fallows, with 37 kilograms per hectare. On uncultivated slopes, phosphate values can be quite high, as in the value of 101 kilograms per hectare. Later in this book, a model will be constructed to explore the consequences of varying fallow lengths on yields. For the purposes of constructing this model, equations have been generated relating fallow

---

16 The unpublished results of Philip Westra (used by permission here) involved 70 sites at the end of the barley harvest in the region of Colta Lake (Map 2); the sites were located between about 3,000 and 3,500 meters and received about 700–900, millimeters mean annual precipitation. The study of Washington Bejarano, José Lainez and Sam Portch, "Adequate Use of Fertilizers on Perennial and Annual Crops in Ecuador," in Elemer Bornemisza and Alfredo Alvarado (eds.), *Soil Management in Tropical America*, Raleigh: North Carolina State University, 1975, 502–517, included five sites to the south of Quito (Pichincha and northern Cotopaxi provinces), from 3,000 to 3,500 meters elevation and with 900 to 1,400 millimeters mean annual precipitation. The two samples taken by the author were from a fertilized and an unfertilized maize field in Imbabura province at 2,800 meters, with a mean annual precipitation of 1,000 mm. All samples were processed in the same laboratory (Santa Catalina). This station uses sodium bicarbonate and disodium EDTA as an extracting solution. A bulk density of Ecuadorian highland soils of 1.1 gram/milliliter was assumed. The conversion factor used to translate meq./100 ml. into kilograms per hectare was 876. These samples were not taken to characterize the mean soil characteristics of the equatorial Andes, but there is no systematic bias either toward soils which are more or less fertile than typical Andean cultivated soils. The samples of irrigated and unirrigated soils at Pimampiro were taken by Patricia Mothes and were processed at the soil laboratory of Texas A&M University. See Mothes, *Pimampiro's Canal*, 195.

17 Where phosphorus levels are less than 22 kilograms per hectare, a yield response to fertilizer application is probable. See S.R. Olsen and L.A. Dean, "Phosphorus," in C.A. Black (ed.), *Methods of Soil Analysis*, Madison: American Society of Agronomy, 1965, 1035–1049. The samples indicated that micronutrients were on average in medium or high supply, except for Manganese (in the sites studied by Westra and Bejarano) and Zinc (Knapp).

18 Fifteen soil samples from the Cayambe plain produced high phosphorus levels (from 20 to 90 kilograms per hectare). See Marcos Wray Espinosa, *Proyecto de Desarollo Ganadero de la Hacienda "Santa Natalia": Pedro Moncayo-Pichincha*, Quito: Facultad de Ingenería Agronómico y Medicina Veterinaria, Universidad Central del Ecuador (Thesis), 1976.

length to initial nitrogen and phosphate, as convenient summaries of limited data.[19]

Environmental Change

The equatorial Andes are subject to frequent earthquakes and volcanic eruptions, and to fairly severe climatic fluctuations.

Several volcanoes have produced ash falls during the last few centuries, including Guagua Pichincha to the west of Quito. The worst destruction, however, appears to have been associated with lahars, especially from the volcano Cotopaxi to the southeast of Quito and northeast of Latacunga (Figure 1.2). Destructive eruptions and lahars affected a wide area between Quito and Latacunga during the years 1742–1768 and in 1877.[20] The widespread falls of ash probably have obscured much evidence for prehistoric and even historic land use, except in the northernmost part of the equatorial Andes.[21]

Earthquakes have been even more frequent. Ambato was destroyed in 1698, Latacunga was destroyed in 1757, and in 1797 an earthquake destroyed the city of Riobamba and stimulated its removal to a location

---

19 For the purposes of constructing this model, the following equation will be used to relate fallow length (r, in years) to initial soil available phosphate (F'), other than that supplied by fertilization: $F' = 21r^{0.4}$. This equation is the product of the fitting of a logarithmic curve to the three data points: fallow 1 year, phosphate 21; fallow 4 years, phosphate 37; and fallow 50 years, phosphate 101. The equation of course has not been demonstrated by the data, but rather provides a convenient summary of the data, for the purposes of constructing a model. Similarly, the following equation will be used to relate initial soil available nitrogen (N') to fallow length in years: $N' = 84r^{0.13}$. The small amount of variation in measured soil nitrogen is reflected in this equation, which is fitted logarithmically to the three data points: 90 kg./ha. at 1 year fallow, 90 kg./ha. at 4 year fallow, and 146 kg./ha. at 50 year fallow. This equation, like the preceding one, will only be used for the purpose of constructing a model of yield response to fallow reduction, and should not be considered as predicting actual soil nitrogen levels.

20 See Minard Hall, *El Volcanismo en el Ecuador*, Quito: Instituto Panamericano de Geografía e Historia, 1977.

21 Ash covers the Formative period Cotocollao and Nueva Era occupation sites. This fact, plus the relative absence of known later sites prior to 600 AD, has led some to infer a catastrophic series of ash falls from 600 B.C. to A.D. 1, followed by perhaps four more centuries of depopulation in much or all of the equatorial Andes and the adjacent Andes of southern Colombia. For this audacious theory see Donald Lathrap, John Isaacson and Colin McEwan, "On the Trail of the Finest Metallurgy of the Ancient New World: How Old is the Classic Quimbaya Style?," *Field Museum of Natural History Bulletin*, 55(10) (1984), 10–19. Lippi, however, has identified a post-Cotocollao occupation which lasted several centuries; see Ronald Lippi, "Paleotopography and Phosphate Analysis of a Buried Jungle Site in Ecuador," *Journal of Field Archaeology*, 15 (1988), 85–97.

farther east. In 1868, the cities of Ibarra, Otavalo, and neighboring towns were destroyed; many irrigation canals were damaged and had to be reconstructed.[22] In 1949, the city of Ambato (Figure 1.2) and neighboring towns were largely destroyed by an earthquake and accompanying landslides.[23] In 1987, a major earthquake inflicted structural damage in the northern highlands, including Ibarra, but was especially severe on the eastern flank of the Andes, where it cut the major trans-Andean oil pipeline.

There is little evidence for much change in temperature in this equatorial mountain environment; mean annual temperatures at Quito from 1891 to 1980 varied only by 2.3 degrees, and mean decadal temperatures by only 1.0 degree. Recent warming might be attributed to the urban heat island effect. There has been considerable variation in frost frequency, however; through 1968, the annual number of frosts observed at Quito has varied from 0 to 45, with an apparent maximum in the 1950s.[24] There is also evidence that the snowline has risen 300 meters or more since the 1500s,[25] although it is unclear which environmental factors may have been responsible for this tendency, and whether there have been corresponding changes in agricultural risk.

The equatorial highlands appear to be insulated from effects of the El Niño torrential downpour phenomenon, which is felt in some years on the coast to the west.[26] The highlands are affected however by "weather regimes" in the northern hemisphere winter, with corresponding relatively dry and relatively wet years.[27] This variation is superimposed on the

---

22 See Teodoro Wolf, *Geografía y Geología del Ecuador*, Quito: Casa de la Cultura Ecuatoriana, 1975, 415–423. Although Quito was partially destroyed a number of times, it apparently never suffered the destruction of the regional capitals; the great bulk of Ecuador's surviving colonial monuments, although not all the archival documents, are in the capital.

23 See Christoph Stadel, "Environmental Stress and Human Activities in the Tropical Andes (Ecuador)," *Revista del Centro Panamericano de Estudios e Investigaciones Geográficas (Quito)*, 15 (1985), 33–50.

24 See manuscript records at the Quito Observatory. Also see Oscar Rovere and Gregory Knapp, "Selection of Climatic Scenarios," in Martin L. Parry, T. R. Carter and N. T. Konijn (eds.), *The Impact of Climatic Variations on Agriculture*, Dordrecht: Kluwer Academic Publishers, 1988, 399–412, and Gregory Knapp, "The Effects of Variations in Mean Temperatures and Frost Risk," in Martin L. Parry, T. R. Carter and N. T. Konijn (eds.), *The Impact of Climatic Variations on Agriculture*, Dordrecht: Kluwer Academic Publishers, 1988, 443–459.

25 Stefan Hastenrath, *The Glaciation of the Ecuadorian Andes*, Rotterdam: A. A. Balkema, 1981, page 58.

26 El Niño of 1982–1983 was an exception.

27 For the concept of weather regime, see Brian Reinhold, "Weather Regimes: The Challenge in Extended Range Forecasting," *Science*, 235 (1987), 437–441. For the argument for the influence of weather regimes on Ecuador, see Knapp, *Soil*, 56–65; Gregory Knapp, "Cambios en el Clima de los Andes Ecuatorianos: Una Tentativa Explicación Genética," in Roy Ryder and Nelson Robelly (eds.), *Geografía y*

"random" interannual variation to produce the observed pattern of moderate interannual variability.[28]

The Crops and Animals

None of the crops grown in late prehistoric times in the equatorial Andes were local domesticates. Rather, they were introduced from the adjacent lowlands or elsewhere in the highlands. Similarly, crops were introduced from Europe after the Spanish conquest. The result has been the deployment of a diverse set of subsistence and commercial crops similar to that found elsewhere in the Andes.

This set of crops includes maize, kidney beans (*Phaseolus vulgaris*), and potatoes, all present prior to the Spanish conquest; and wheat, barley, broad beans (*Vicia fava*), sugar, and peas, all introduced by the Spanish in the sixteenth century. In 1985, fully 90 per cent of the area sown in the three northern provinces of the equatorial Andes was in one of these eight crops.[29] These crops today provide the bulk of the calories in the highland diet, although rice imported from the lowlands is gaining importance.[30]

The major crops include numerous varieties or land races, with different aesthetic and culinary characteristics.[31] Karl Zimmerer's yield data from southern Peru does not support the notion of land race adaptation to specific environmental niches. Andean land races rather represent the

---

Desarrollo: Metodologías y Casos de Estudio, Quito: Instituto Geográfico Militar, 1982, 19–26.

28 The longest record, for Quito, shows a range between 692 and 1,905 millimeters for the years between 1891 and 1983.

29 Carchi, Imbabura, and Pichincha provinces (less Santo Domingo cantón), excluding those crops which are grown in the tropical lower elevations. See INEC-MAG (Instituto Nacional de Estadística y Censos – Ministerio de Agricultura), *Sistema Estadística Agropecuario Nacional por Muestreo por Areas*, Quito, 1985.

30 The eight crops (not including rice) accounted for about 76 percent of the calories consumed in the late 1950s at Tocachi, 58 percent at San Rafael, 75 percent at Malchinguí, 55 percent at Pomasquí, 65 percent at Tambillo, 69 percent at Pifo, 71 percent at Aloag, 58 percent at Cumbaya, 77 percent at Alangasí, and about 70 percent at Peguche. See María Vargas and Sylvia Gallegos, *Diagnóstico Nutricional y Alimentario del Ecuador*, Riobamba: Escuela Superior Politécnica de Chimborazo, no date.

31 Maize varieties can be divided into hard corn types, mostly grown in tropical low elevations and typically used as animal feed, and soft corn types, the most commonly found in the highlands. Popcorn (*canguil*) is a special and common crop in the highlands. Maize varieties include *chillo*, *guandango*, *precoz*, *mishca*, *chaucho*, *chulpe*, *morocho*, and *morochillo*. Potato varieties include *chola* (*yungara*), *violeta*, *guantiba*, *pintada*, *moronga*, *runamayo*, and *papa chaucha*.

absence of selection for uniformity in a mountain landscape where seed has traditionally been obtained locally.[32]

The major crops are important both nutritionally, and as components of a rich and complex highland cuisine. The major dishes are soups and stews, containing pulses (lentils, peas, beans) , grains (barley, maize) and chenopods, vegetables (squash, greens), and/or tubers (potato, oka, melluku) in a bewildering variety of combinations. Finger foods are also important, including toasted foods (maize, toasted or popped; tawri, broad beans). Sugar cane is eaten as a delicacy or purchased as unrefined sugar.[33]

This Andean cuisine is unusually flexible; food items may be added, subtracted, or substituted from soups, stews, and finger foods, and no single crop is indispensable. Unlike much of the world, where a single culturally given staple (maize in Mexico, rice in Japan) dominates the cuisine and the landscape, Andean farmers can choose between a number of almost equally acceptable crops. Relative productive efficiency (in calories per unit labor) can thus play some role in the choice of the major staple.[34]

The needs for fiber were met in late prehistoric times by cotton and camelid wool production, as well as cabuya fiber; today only sheep wool is a major product.[35] Meat in the pre-Columbian period was provided by the native household animal, the guinea pig; since the Spanish conquest, chickens, goats, cattle, and pigs have provided meat and milk which have played an increasing role in the local diet. Animals are fed with kitchen refuse, or are pastured in high grasslands, meadows, and fallow fields. With the partial exception of maize (maize stalks), there is little tradition of growing fodder crops in the equatorial Andes. Nor is there a tradition of transhumance, in a context where pastures are available year round, nor of pastoralism, since pasture area is restricted relative to crop land.

---

32 See Zimmerer, *Seeds*.

33 A set of soup and stew recipes can be found in Ralph Beals, *Community in Transition: Nayón-Ecuador*, Los Angeles: University of California at Los Angeles, Latin American Studies Volume 2, 1966. A useful study of Ecuadorian cuisine is Mary Weismantel, *Food, Gender and Poverty in the Ecuadorian Andes*, Philadelphia: University of Pennsylvania Press, 1988.

34 Kathleen Schroeder, "Señora Córdova's Kitchen: Food Substitution in the Sierra of Piura," Annual Meeting of the Association of American Geographers, Toronto, Canada, 1990.

35 Wool in prehistoric times would have been from llamas; since the Spanish conquest sheep have played this role.

Methods for Identifying the Strategically Relevant Environment

It is hardly surprising that reliable information on crop yields in the Andes is difficult to obtain; such information is scarce for subsistence farming areas around the world. Nevertheless, without some attempt to collect such information, any attempt to determine the relevant environment for Andean peoples is arbitrary. Here several data sources and methods will be explored: modelling, perception, interviews, and several kinds of production records and trials, including hacienda records, government experimental farm records, and national statistics.

CROP-CLIMATE MODELLING

Perhaps the first major effort to model the relationships between crop yield and climate in this region is that attempted by Andrés Ravelo and his team, for the specific case of barley.[36] His work shows considerable promise, but there are also formidable obstacles. Perhaps foremost is the fact that the technique of crop-climate modelling, as developed in the northern hemisphere, assumes optimal fertilization. As will be seen below, fertilization is normally far from optimal in the Andes. The aforementioned study did produce the suggestive result that there is an optimal precipitation value for barley yields,with yields decreasing in both wet and dry years.[37]

PERCEPTION

There has been a surprising scarcity of efforts to ask Andean farmers themselves about the significant environment for farming. Perhaps the old presumption of scientific superiority to Andean peasant perception in some cases was replaced by an assumed pan-Andean environmental perception ideal, constant since Inca times. In 1980–1981, a farmer near the Colta Lake, assigned to the author by the local indigenous association, was asked to list significant environmental niches and events for himself and his neighbors; his detailed description was cross-checked against several other farmers.[38] The results of this inquiry stimulated a later, independent study

---

36 See Andrés Ravelo, Ana M. Planchuelo-Ravelo and Tom Hodges, "The Effects of Climatic Variations and Agricultural Practices on Barley Yields," in M. L. Parry, T. Carter and N. T. Konijn (eds.), *The Impact of Climatic Variations on Agriculture. Volume 2: Assessments in Semi-Arid Regions*, Dordrecht, Netherlands: Kluwer Academic Publishers, 1988, 429–442.

37 The optimal value was to be found between moderately wet and extremely wet years, or between 451 and 1,744 millimeters of rainfall in the October–June growing season.

38 Knapp, *Soil*, 149–150.

near Guamote, involving a larger sample.[39] Further information was gathered from a few informants at Zuleta and San Pablo near Otavalo in the northern highlands.

The results of these surveys, although preliminary, may be summarized as follows. When asked about the desirable zones for farming, indigenous farmers respond not with altitudinal zones[40] but with a more complex topographical, edaphic and climatic terminology.

The most desirable environmental niche for potato or maize farming is called *pugru*; the reference is to a topographic low point, a valley bottom or swale (*ondonado*, *bajo*, *bajío*, *hondón*). Here soils are considered to be deepest, water supply greatest, and farmers perceive heightened protection from frost and disease.

Flats (*pamba* or *pampa* in the more standardized Quichua pronunciation), one kind of pugru, are considered prone to frost but have readily irrigable soils for productive pasture. Thus flats are normally the most expensive land; near Lake Colta, flats were worth $5000 (US) per hectare in 1981, three times the value of the adjacent slopes.[41]

In the northern highlands, the term *rinconada* refers to the flat upper end of a U shaped glacial valley, floored with glacial outwash. These are coveted by farmers.[42]

Slopes (*quingre*, *kiri*, or *kinri*) are another category, with little frost risk but with frequently shallow soils.

Soils (*allpa*) may be distinguished as good (*alli allpa*) or as bad (*mana allpa*). The best soil is thick, black prairie soil (*yana allpa* or *yanallpa*), full of earthworm casts (*kuika*), *juzu* grubs, and *juzu* grub droppings. A black soil with greater clay content, representing an exposed subsurface horizon with clay and humus accumulation, is called *gridallpa* near Colta; in

39 Luis Cañadas, Tarquino Salcedo, Raul Bravo and Gregory Knapp, "The Vulnerability of Indigenous Farming Systems to Climatic Variations," in Martin L. Parry, T. R. Carter and N. T. Konijn (ed.), *The Impact of Climatic Variations on Agriculture. Volume 2: Assessments in Semi-Arid Regions*, Dordrecht: Kluwer Academic Publishers, 1988, 416–417. My role in this study was not in the conducting of interviews but rather in the interpretation and analysis of the results.

40 There is of course a recognition that some land is in páramo grass; this land is called *urku* (*jagua urku*, *jatun urku*). Although it may imply an altitudinal zone, that is, land which is too high to cultivate, some farmers at least use the term to refer to grassy land which could be cultivated. Cultivated land can be referred to as *uku allpa* (literally, inside land), *tarpuna allpa* (land for sowing), *anejos* (settlements; see Alfredo Costales and Piedad Peñaherrera de Costales, *Katekil o Historia Cultural del Campesinado del Chimborazo. Llajta No. 4*, Quito: Instituto Ecuatoriano de Geografía e Historia, 1957, 37–38), or *tierra baja*. The west slope of the Andes below the páramo is called *sacha* (forest); the lower forest is called *yunga*.

41 See Gonzalo Rubio Orbe, *Punyaro*, Quito: Casa de la Cultura Ecuatoriana, 1956, 109, for the case of Imbabura.

42 The verb *rinconadarse* is used to refer to the post-Conquest upslope movement of farmers after the Spanish conquest.

Imbabura it may be called *chocoto* or *barro*. This soil is considered suitable for agriculture and brickmaking. Another common soil is the exposed cemented hardpan (carbonate or silica cementation) commonly known among Spanish (and some Quichua) speakers as *cangahua* or *cancahua*. In Chimborazo, this soil is called *killin* (*killin allpa, killin rumi, killu allpa sinchi*). Various kinds of eroded soil are recognized: sandy soil (*tiu allpa*), wind-eroded soil (*guaira allpa*), thin soil (*tullu allpa*) and stony soil (*rumi allpa*).

Soils are not perceived as static but rather the results of human activity. Some of the informants reported cases of soils losing their top horizon within 5 to 50 years of cultivation and thus changing from *yanallpa* to *killin*; others also recalled similar stories from parents or grandparents. Alternatively, farmers near Lake Colta have transformed exposed hardpan into valuable soil by hacking it into bits with a pickaxe, allowing the chunks to slake in the rain, pulverizing the wet clods, and mixing with manure.[43]

All of these terms are of special importance to farmers with respect to crop yields. Farmers realize that fresh lands, reclaimed from grassland or forest, yield the most; around Colta, one informant estimated that fresh land yielded up to 13,000 kg/ha of potatoes, with a drop to 7,000 kg/ha in the second year and 5,500 kg/ha in the third. Another source indicated that in good years, *pugru* land with *yanallpa* could yield up to 13,500 kg/ha, while adjacent steeper slopes might be expected to produce 7,200 kg/ha.

These figures suggest that soil quality and position are perceived to affect yields by fifty per cent or more. In addition, farmers are well aware of other hazards. Frosts (*rasu, k'asa*) are considered especially significant. At San Pablo, frosts are named after the nearest religious holiday, and the most damaging ones are the ones in January–February (epiphany frosts, *heladas de Santos Reyes*; Ash Wednesday frosts, *heladas de ceniza*), and March–April (Palm Sunday frosts, *heladas de ramas*). In Chimborazo, informants suggested that the most damaging frosts for barley, potatoes, and broad beans occurred in the May–June period, with a second hazard peak in November–December. Frosts are actually most common in the summer dry season (June–September) (*heladas generales*), but these are not considered especially dangerous since there are few growing crops at this time.

In both Imbabura and Chimborazo, it was clear that there were relatively few strategies effective against frost; farmers frequently professed total ignorance on the point. Farmers perceived the most important strategies to be staggered planting dates (*suertes*), the use of slopes rather than flats, and the use of east-facing slopes to catch the morning sun. Some farmers perceived that ridging (*wachukuna*) limited frost damage. Near San Pablo, some of the non-Indian farmers with large plots of land burned heaps of straw early in the morning to avert frost damage.

Drought (*usiag, chaquishca, sequía*) is also often mentioned as a danger. In San Pablo, the late 1970s and the year 1927 are remembered as particularly difficult periods. The importance of rain is indicated by the

43 See also Beals, *Community*, 19–20.

variety of methods used to predict its occurrence. In summer, noises from the volcano Tungurahua and the descent of foxes (locally called *lobos*) from the higher elevations are supposed to indicate rain. An earthquake is supposed to change the prevailing pattern to its opposite: wet to dry, dry to wet. Many farmers also rely on the Bristol Almanac (Spanish edition), which contains climate predictions and is sold on market days in towns such as Guamote.

Excessive rain (*tamia sinchi*, *aguacero*) is an equally important perceived problem. Many slopes and most flats have a complex artificial drainage system intended to channel the initial surface runoff from storms and prevent surface sheetwash; in some cases farmers actually create planting ridges parallel to the slope to speed the runoff process.

Winds are not considered a major problem. However, during dry periods, whirlwinds (*huracanes*) travel over the bare fields, removing soil and organic matter. A common response to this is working of the soil and spread of a grass mulch to limit wind erosion of seed, soil, and fertilizer (*shigua paja*).

The most important potato disease is called *lancha*, a fungus infestation (*Phithoptera infestans*). The same word "lancha" is however used for a variety of other conditions, including diseases of maize. *Lancha* is often attributed to weather conditions: a combination of rain and bright sunlight, or bad "air" which comes at night. Grubs (*gusano blanco*, *juzu*) are also considered to pose problems for potatoes, while *la roya* is a disease of barley.

INTERVIEWS FOR QUANTITATIVE DATA

Farmers were interviewed to provide quantitative data on the relationship between climatic parameters and yield. In the relatively dry communities near Guamote in southern Chimborazo province, questionnaire results from 50 farmers indicated that in a "wet" year (about 650 mm October–June precipitation), potatoes yielded 4.2 metric tons/hectare, while in a "dry" year (300–400 mm October–June precipitation), potatoes yielded 1.2 metric tons/hectare, assuming the application of organic fertilizer. Similar figures for barley were 826 kilograms/hectare in a wet year, and 335 kilograms per hectare in a dry year. Without fertilizer, yields in either kind of year would be below seed replacement.[44]

44 See Cañadas, Salcedo, Bravo, and Knapp, "Vulnerability," 424–426.

Agricultural Censuses and Sampling

Agricultural censuses and sampling procedures produce data on yields which are of relatively little use to the study of yield-environment relationships, because the regional units of reporting are too large and include widely varying environments. The census values are however useful as indicating averaged overall potato and maize yields.

Maps and Field Correlations

Land use maps provide information on the empirical relation of crops and crop zones with environmental parameters, of course without revealing the underlying processes responsible for these patterns. These data are especially useful for identifying the upper and lower limits of cultivation of crops.

The equatorial Andes have been mapped by Gondard and López, and these maps reveal certain ecological patterns in the key crops and their zones of cultivation (*paisajes*). The absolute upper limit of agriculture in the equatorial Andes (potatoes) is at about 4,000 meters, but fields are rare at such elevations. The upper limit of potatoes in Imbabura province is at 3,600 meters, with only 8 per cent of the potato area above 3,200 meters. In Chimborazo Province, the upper limit of potato cultivation is at 3,800 meters, with about 8 per cent of arable land between 3,600 and 3,800 meters.[45]

The upper limit of potato cultivation in flats is lower than the upper limit on slopes; one of the highest intensively cultivated flats in Ecuador is at Llano Grande, Chimborazo, at 3,080 meters. The small cultivated fields around Lake Colta at 3,400 meters may be the highest on flats in the country.[46]

Maize usually has a lower upper limit of cultivation than potatoes. In Imbabura Province, it is cultivated commercially at least up to 3,100 meters and on a small scale at higher elevations. Its upper limit, on slopes around Lake Colta, is about 3,400 meters.

---

45 The land use maps have been published under the titles PRONAREG-ORSTOM (Programa Nacional de Regionalizaci6n Agraria - Office de la Recherche Scientifique et Technique Outre-Mer), *Mapa de Uso Actual del Suelo y Formaciones Vegetales, 1:200,000*, Quito: Instituto Geográfico Militar, 1983. I have consulted the sheets for Quito, Ibarra , Tulcán, and Riobamba. The methodology for the maps has been described by Pierre Gondard, *Inventario y Cartografía del Uso Actual del Suelo en los Andes Ecuatorianos*, Quito: MAG, PRONAREG, ORSTOM, and Museo del Banco Central, 1984. See the ingenious diagrams relating cultivated areas of potatoes, maize, and sugarcane to altitude in Gondard and L6pez, *Inventario Arqueol6gico*, 35–36.

46 Knapp, "Effects," 451.

A number of crops are not found above about 2,500 meters. Sweet potatoes and sugar cane, and probably cotton, are found up to about 2,500 meters, avocados up to 2,400 meters, and manioc to only 1,800 meters. There are also observable precipitation limits to unirrigated crop zones. In Chimborazo Province, the area under cultivation falls almost entirely within the area with from 400 to 1,400 millimeters mean annual precipitation, with only barley being suitable for areas near the lower limit.[47] These limits appear to hold elsewhere in the equatorial Andes, with most crops having narrower ranges.

The lower precipitation limit of maize is of special interest. Land use maps and field observations suggest a lower limit of maize cultivation about 600 millimeters mean annual precipitation.[48]

These observed limits indicate empirical correlations between crop yields and environmental factors. Further exploration of these correlations requires the use of agronomic trials. These trials can show the relative significance of soil fertility, temperature, precipitation, and slope for crop yields and thus more precisely delineate the strategically relevant environment.

PRODUCTION RECORDS AND TRIALS

Soil fertility is dependent on numerous soil characteristics, and is relative to the specific needs of the crop plant. This complexity however has often resulted in soil being ignored as a critical component of crop success. At the risk of oversimplification, an attempt will be made here to devise a general soil fertility index for the equatorial Andes, which can be used to determine the relative roles of soil fertility and other factors in crop yields. This soil fertility index will be used extensively in the rest of this book.

Fertilization experiments were carried out with potatoes on volcanic ash soils in southern Pichincha and northern Chimborazo provinces, during the agricultural year 1972–1973. Table 2.2 shows the results of these trials in higher elevation areas (3,000–3,500 meters).

---

47 See Luis Cañadas, Trajano Yugcha and Washington Estrada, "The Effects on Zones of Agroclimatic Suitability for Crops," in M.L. Parry, T. Carter and N. T. Konijn (eds.), *The Impact of Climatic Variations on Agriculture, Volume 2: Assessments in Semi-Arid Regions*, Dordrecht: Kluwer Academic Publishers, 1988, 461–471.

48 This is based on interpolations between stations. Patricia Mothes was able to locate the "first" maize field to use irrigation in the Pimampiro valley; it is located about 55 percent of the way between the stations of Mariano Acosta and Pimampiro, suggesting an interpolated precipitation of about 750 mm per year; see Mothes, *Pimampiro's Canal*, 31–37, 129. On the route from Ibarra to the Chota Valley, the last maize field without irrigation is located about at the 600 mm isohyet; conversations with the farmer established that this was a speculative field, with losses in some years and good yields in others.

TABLE 2.2 YIELD AND FERTILITY DATA, BEJARANO

	Locality			
	STA. ROSA	UMBRIA	STA. CATALINA	AYCHAPICHO
Initial phosphate (kg./ha.)	41	37	64	18
Initial nitrogen (kg./ha.)	68	116	82	82
Limiting nutrients	P, Mn	P	P, Mn	P
Yields as per cent of maximum (with N-P2O5 fertilizer added):				
0–0, yield as percent of maximum	43	43	55	29
0–200, yield as percent of maximum	72	82	90	87
Maximum yield (0–400) (tons/ha)	*16.95*	*22.94*	*13.54*	*27.15*
0–200, yield as percent of maximum	42	45	54	55
50–200, yield as percent of maximum	79	84	93	78
Maximum yield (200–200) (tons/ha)	28.95	41.78	22.47	42.54

Expressed as a percentage of optimum yield, response to soil fertility was remarkably similar in the four sites; it appears that approximately 100 kg. nitrogen and 200 kg. added phosphate per hectare results in excellent yields in a variety of contexts.

It is possible to use the data provided here to calibrate a simple soil fertility index for the equatorial Andes. The difficulty of calculating a fertility index resides in the twin facts that (1) a number of different nutrients contribute to soil fertility and (2) available nutrients are a sum of nutrients available prior to fertilization and those supplied by fertilizer. In the equatorial Andes, however, the problem is simplified by the relative absence of nutrient deficiencies except for phosphorus, and broad similarities in the pre-fertilization status of soils.

Willcox[49] and Bear[50] have suggested that the contribution of individual nutrients to yield can be approximated by an equation of the form:

$$Y(X) = a(1-10^{-cX})$$

where Y is predicted yield as a function of nutrient X, a is the maximum yield obtainable by manipulating supply of nutrient X alone, X is the available supply of nutrient X, and c is a constant partially dependent on crop and local soil conditions.[51]

For practical purposes, X can be considered as consisting of three components: X = x' + x" + x"', where x' is the measured quantity of the available nutrient in the soil, x" is the quantity of the nutrient applied in fertilizer, and x"' is the unmeasured supply of the nutrient (released during the crop cycle from organic matter, delivered by atmospheric dust or irrigation, etc.). The above equation can be rewritten as:

$$Y(X) = a\ (1- 10^{c(x'+x''+x''')} \quad \text{or}$$

$$1-(Y(X)/a) = b \cdot 10^{c(x'+x'')} \text{ where}$$

$$b = 10^{cx'''}$$

The last form of the equation is suitable for approximation by an exponential regression. If we perform such a regression for yield response to phosphate (F), using those yields with no added nitrogen and 0 or 200 kg/ha added phosphorus (Table 2.2), we obtain the following equation[52]:

$$Y(F) = a\ (1- .726 \cdot 10^{-.0027(f'+f'')})$$

Here f' and f" are phosphate in the soil and in the fertilizer respectively. The multiplier of .726 is equivalent to 51 kg/ha of unmeasured phosphate. For nitrogen (N), examination of Table 2.2 shows scant response to differences in measured soil nitrogen in the plots not fertilized with this nutrient. This is not surprising, given the difficulty of adequately measuring soil nitrogen and the role of soil organic matter in delivering nitrogen throughout the soil cycle. Therefore, here the regression was performed without including an estimate for soil nitrogen.[53] The resulting

49 See O. W. Willcox, *ABC of Agrobiology*, New York: Norton, 1937; O.W. Willcox, "Verification of the Mitscherlich Effect Law," *Agronomy Journal*, 43 (1949), 83–90, and O. W. Willcox, "Quantitative Agrobiologic Evaluation of Some Phosphate Fertilizer Trials with Vegetable Crops in Alabama," *Agronomy Journal*, 43 (1951), 83–90.

50 See Firman E. Bear, *Soils in Relation to Crop Growth*, New York: Reinhold, 1965.

51 Willcox, following work done in Germany before World War II, felt that c was a constant regardless of crop and soil conditions, although this is a view which has fallen into disfavor; see Roger H. Bray, "Correlation of Soil Tests with Crop Response to Added Fertilizers and with Fertilizer Requirement," in H. B. Kitchen (ed.), *Diagnostic Techniques for Soils and Crops*, Washington D.C.: American Potash Institute, 1948, 53–86.

52 The regression on eight data points produced an r square of .856.

53 The regression on eight data points produced an r square of .748.

equation was:

$$Y(N) = a\,(1 - .506 \cdot 10^{-.011(n'')})$$

Here n'' is the quantity of fertilizer nitrogen alone. The value of b (.506) suggests an unmeasured nitrogen component (including soil nitrogen) equivalent to 27 kg/ha. For two critical nutrients X_1, and X_2, yield can be predicted as the product of yield equations:[54]

$$Y(X_1, X_2)/a = Y(X_1)\,Y(X_2)$$

For the case of nitrogen and phosphate,

$$Y(N,F)/a = (1 - .506 \cdot 10^{-.011(n'')})(1 - .726 \cdot 10^{-.0027(f'+f'')})$$

The value of this equation varies from 0 to 1 depending on the quantity of fertilizer applied, and signifies the contribution of nutrients to yield. We will call it the "fertility index" in the rest of this book,[55] and abbreviate it with the letter I:

$$I = Y(N,F)/a.$$

This equation was calibrated for potatoes in the specific sites of the study, but will be used here as a rough index of soil fertility for both maize and potatoes in the equatorial Andes. As with all generalizations, the justification for this procedure lies with its heuristic value in the further exploration of environment-yield relationships, as seen in the next section[56]

The Galo Plaza Haciendas

Production data have been made available from several large farms owned by the late Mr. Galo Plaza Lasso in the area between Lake San Pablo and Zuleta (Figures 1.2, 2.1, 2.2): Haciendas Angla, Topo, and Zuleta. In the following pages, for brevity, these haciendas will be referred to as the "Plaza farms." Individual fields (uniform management areas) of these farms are labelled with capital letters on Figure 2.1. The fields are large, and may span several hundred meters from their upper to their lower ends. In each case, an elevation of the midpoint of the field was determined, and a

---

54 In modern fertilizer tests quadratic equations are preferred to express yield since massive nutrient doses can actually decrease yield. For our purposes, however, the exponential equations are simpler and easier to use, and adequately express decreasing returns as long as nutrients are not allowed to be present in excessive quantities, either absolutely or in relation to other nutrients.

55 The fertility index used here is calculated somewhat differently than in earlier publications by the author, and therefore is a different function of F and N. This is due to the fact that the regressions used to derive values for the coefficients b and c here omit the more heavily fertilized trials. The differing index calibration does not affect the major conclusions of this study.

56 The problems of defining soil fertility in quantitative terms have been partly responsible for the dearth of studies which empirically study soil fertility in relation to real yields. The development of regional soil fertility indexes which are applicable to varied crops may be of limited utility to agronomy but are of potentially great value for historical cultural ecology.

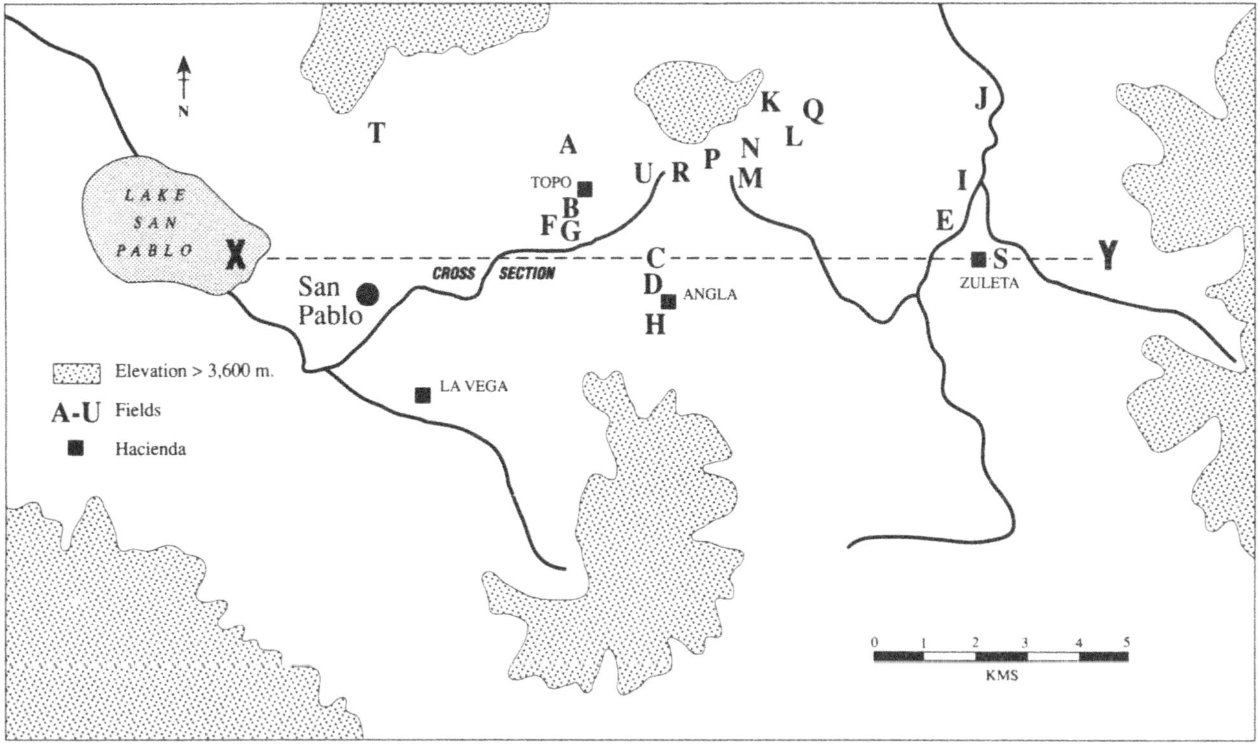

Figure 2.1 Location of Fields, Plaza Farms

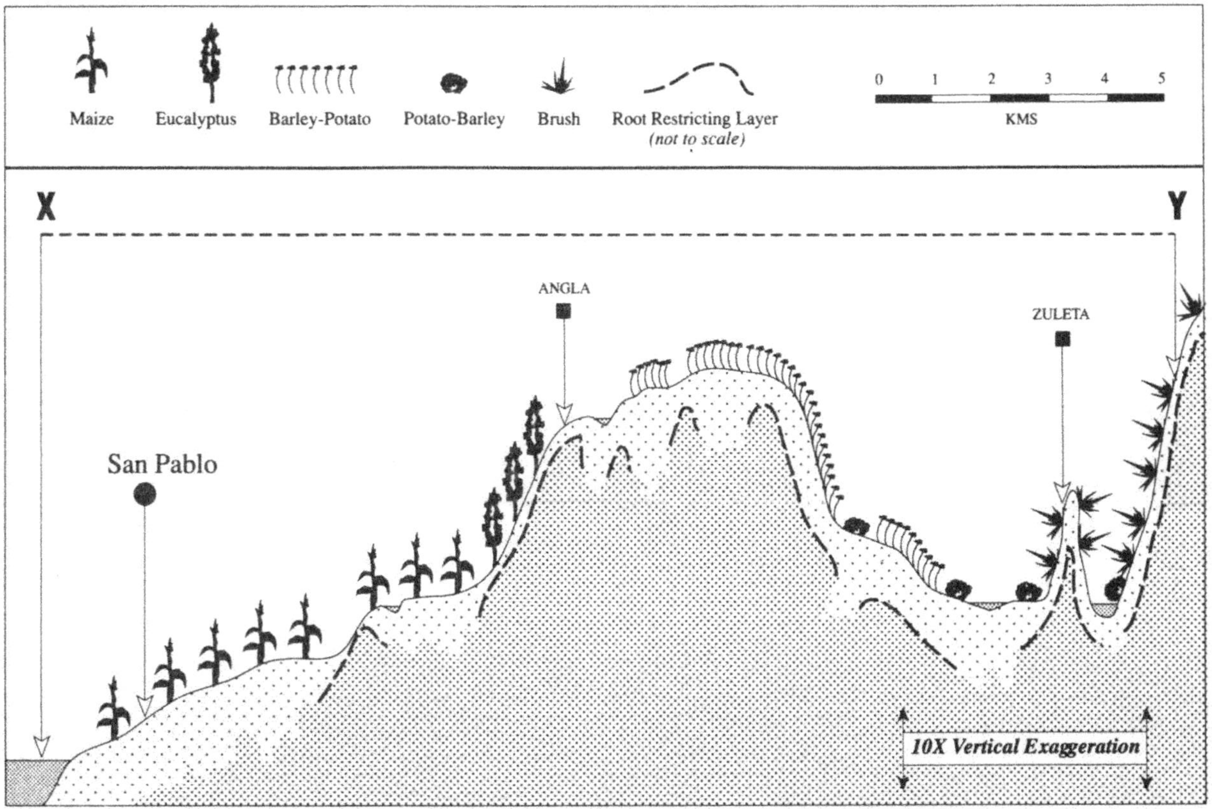

Figure 2.2 Transect, Plaza Farms

median slope value estimated on the basis of topographic maps. Because this volcanic landscape is marked by gradual rather than abrupt topographic changes, the elevation and slope values for these large fields are reasonable generalizations.

The fields are conveniently located near precipitation stations (at Angla, Zuleta, and San Pablo on Figure 2.1); they receive about 1,000 millimeters per year of average rainfall.[57]

Elevation of the fields vary from 2,800 to 3,500 meters, while slopes vary from flats to 35 per cent. Figure 2.2 is a cross section of the terrain from point X to point Y in Figure 2.1. Land use and depth to root-restricting layer are illustrated.[58] The soils of the Plaza farm fields are developed on volcanic material, varying from deep fertile soils in the flats to shallow soils on the steeper slopes.

The crops of the Plaza farms received a minimum of topographic or microclimatic control, and were partially protected from pests and drought[59] They were thus more likely to reflect environmental influences of slope, excess precipitation, frost, or fertility. These large-farm yields may be helpful therefore in the derivation of simple productivity equations which may then be applied—in suitably altered form—to the more complex case of small-farm technology.

From 1965 to 1979, the fields were planted to different crops, and data on quantity of seed, fertilizer, pesticides, and yield were recorded in the

---

57 For the 1964–1973 decade, precipitation at San Pablo averaged 1,039 millimeters, for Angla 970 millimeters, and for Zuleta averaged 1,028 millimeters. The data are derived from INAMHI meteorological annuals and manuscript lists at PRONAREG-ORSTOM, Quito.

58 Based on PRONAREG-ORSTOM, *Mapas de Suelos*, and manuscript land use maps and soil profile descriptions of PRONAREG-ORSTOM, used by permission.

59 The following notes as to hacienda cropping practices may help in interpreting the yields in the tables. Potatoes are grown in ridges (*wachukuna*), perpendicular to the slope (left-right, rather than up and down). The land is worked with plows, usually including two or three weeding/ridging operations (*aporques*). Potatoes are fertilized with a variety of chemical fertilizers and are treated with fungicide. There is no evidence in the records of severe pest loss except perhaps in time of frost. Fields in flats may be occasionally irrigated. Potatoes are normally not intercropped but they are rotated, especially with wheat and barley. Pasturing is part of the sequence, involving the deposition of manure. In flats. grass seed (orchard grass or Italian ryegrass) is often planted after the potato/small grain sequence, allowing grazing on improved pasture until the inevitable invasion by *kikuyu* grass (*Pennisetum clandestinum*). Maize is grown in ridges similar to those for potatoes, and is usually harvested fully dried and ripened. Maize is usually grown alone, although a record of a sharecropping harvest showed that intercropped pulses equalled 11 percent of maize production by weight. Hacienda maize fields are often rotated with wheat and/or rested for several years between maize harvests; they are also usually fertilized with chemical fertilizers, but pesticides are not normally added. The result is that 62 percent of the harvests are listed as affected by rotting (*pudrición*) at least to some degree. In this chapter we will eliminate rotten maize from the harvest totals.

farm record books.[60] The record books give yields in terms of return to seed, which can however be converted to yields per hectare. [61]

A number of varieties of maize and potatoes were planted by the Plaza farms. The traditional Yungara variety of potato has recently been supplemented by a number of new varieties with higher yields. For this study we will use yield data only from the traditional Yungara variety.[62]

Several varieties of maize are also planted; the small size of the data set required amalgamation of Chillo and Santa Catalina yields for analysis.[63]

TABLE 2.3 POTATO YIELDS ON FLATS (ALL VARIETIES), PLAZA FARMS

Planting	Year	Yield (kg./ha.)
8	1970–71	5.8
10	1972	24.8
23A	1976–1977	8.9
23B	1976–1977	0.0*
23C	1977	16.6
23D	1977	18.4
27	1978–1979	6.7
17	1974–1975	0.0*

*Frost

The data on yields for the chosen potato and maize varieties are shown in Tables 2.3, 2.4, and 2.5. Each planting corresponds to a particular field

60 The record books do not record date of planting. It is assumed here that potatoes planted and harvested in the same year were planted in June–July and harvested in November–December. For potatoes harvested in the year after planting, it is assumed that in areas under 3,050 meters, the planting was in September–October and the the harvest was in February–March; while in areas above 3,350 meters, planting occurred from the end of June to September and harvested from the end of February to the beginning of June. These assumptions, based in part on interviews and in part on regressions of precipitation against yield, are necessary for the calculation of mid growing season precipitation.

61 According to interviews, 20 to 24 *quintales* are of potatoes are planted per hectare (910 to 1,090 kilograms) we will assume a constant value of 1,000 kilograms. Maize is planted at the rate of 36 kilograms per hectare.

62 This assumption was required by expediency. It has since been demonstrated in southern Peru that groups of cultivars belonging to the same potato species are not adapted to distinctive altitudinal environments. See Zimmerer, *Seeds*, 275.

63 Varieties planted included the traditional *chillo*, *morocho*, and *canguil* (popcorn), as well as *mishka*, *chulpe*, and Santa Catalina.

TABLE 2.4 PLAZA FARMS YIELDS: POTATOES

Planting, potatoes	Site	Year	Area (ha.)	Soil	Slope (%)	Elev. (m.)	Mean Precip. per month (mm.)	Yield (kg./ha.)
1	I	66–67	12.3	—	10	2850	153	1544
2	D	67	3.5	R	12	3050	54	5950
3	J	68	6.9	—	15	2800	86	13810
4	K	68–69	10.7	R	35	3500	94	9150
5	L	68–69	10.3	R/S	15	3500	94	9987
6	M	69–70	10.5	R/S	15	3350	120	10724
7	K	70–71	17.6	R	35	3350	128	7955
12	N	72–73	8.2	R	35	3500	77	14067
13	P	73–74	10.8	R	35	3450	94	14343
14	Q	73–74	3.9	S	15	3350	154	2682
15	J	74	5.4	—	15	2800	97	17586
16	R	74–75	9.0	R	25	3450	119	16062
17	S	74–75	0.2	—	Flat	2850	208	0
18	I	75	2.0	—	10	2850	90	13418
19	B	75	0.2	—	12	2950	48	0
20	R	75–76	5.6	R	25	3450	123	14891
21	R	76–77	3.1	R	25	3450	65	18735
22	R	76–77	3.8	R	25	3450	65	20131
24	R	77–78	3.6	R	25	3450	68	15987
26	T	77–78	3.2	S	25	3000	82	11577
29	U	78–79	5.0	R	20	3050	74	11518

TABLE 2.5 PLAZA FARMS YIELDS: MAIZE

Maize planting	Site	Years	Area (ha.)	Soil type	Slope (%)	Elev. (m.)	Mean precip per month (mm.)	Yield (kg./ha.)
1	A	65–66	1.3	—	12	3060	77	1667
2	B	66–67	9.3	—	12	2950	103	1127
3	F	67–68	11.1	—	8	2900	85	1228
4	B	68–69	13.3	—	12	2950	87	2023
6	C	70–71	3.8	R	20	3100	103	1166
8	C	71–72	2.5	R	20	3100	99	490
9	D	72–73	2.5	R	12	3050	52	4540
10	D	73–74	2.7	R	12	3050	102	3290
11	D	74–75	4.2	R	12	3050	89	2930
12	B	75–76	0.3	—	12	2950	82	1282
13	H	75–76	7.9	R	12	3050	82	2574
14	B	76–77	0.3	—	12	2950	55	0
15	E	76–77	8.9	—	12	2850	94	3006
16	F	77–78	14.2	—	8	2900	93	1908
17	G	78–79	5.3	—	8	2900	94	1040

or site (located on Figure 2.1) and a particular year. Data on slope[64], elevation of midpoint of field, and soil problems are derived from soil maps[65] and topographic maps. Yield and fertilizer application data are derived from hacienda figures.[66]

The two soil problems listed in the tables are shallow depth to root restricting layer (labelled R) and sandiness or stoniness (labelled S). Both problems may be expected to reduce soil moisture and nutrients available to plants.

---

64 Only one of the fields is mapped by PRONAREG-ORSTOM as having a slope of under 5 percent (by our definition, flat). Other fields have flat portions too, however, and, our figures for slope should be considered indicative only of preponderant conditions.

65 See PRONAREG-ORSTOM, *Mapas de suelos.*

66 Yields are expressed in hacienda books in terms of yields per unit seed. Fertilizer is also expressed in terms of application per unit seed. Both are converted in terms of unit area, based on the normal seed/area conversions mentioned above. Fertilizer data for phosphorus is expressed as phosphate (P_2O_5) equivalent. Potassium is not shown as it is normally not a limiting factor in Ecuadorian volcanic soils; see Bejarano, Lainez and Portch, "Adequate Use."

Figures for precipitation per month during the middle of the growing season have been listed in the tables.[67]

The yield data will first be discussed in terms of losses to frost, then in terms of fertility, elevation, and precipitation.

PLAZA FARMS: FROST

Two of eight potato plantings in the flats (of all potato plantings, including non-Yungara varieties) were recorded as lost to frost in the records (plantings 17 and 23B, Table 2.3). Two out of five plantings in flats were lost to frost out of those planted and harvested in distinct years. None of the three plantings planted and harvested in the same year were lost. This suggests a 40 per cent risk of loss for late year plantings in flats.

Only one in twenty Yungara plantings on slopes (planting 20, Table 2.4) were lost to frost. Frost risk is greater on flats than on slopes. None of the maize plantings on slopes or flats was recorded as lost to frost.
In the following steps of analysis of yields, the two crops lost to frost will not be included, though it must be borne in mind that parts of other harvests may have been lost to frost.

PLAZA FARMS: FERTILITY

Inspection of Tables 2.4 and 2.5 suggest that fertilization may be a major factor in accounting for variation of yield. For potatoes, the one unfertilized plot yielded 5,900 kilograms/hectare; heavily fertilized plots

---

67 The study by J. Glover, "Relationship Between Total Seasonal Rainfall and Yield of Maize in the Kenya Highlands," *Journal of Agricultural Science*, 49 (1957), 285–290, indicated that yield in tropical Africa was best predicted by the precipitation of the period beginning one month after sowing and ending one month before harvest—about 70 percent of the total growing period. On the same principle, we will use precipitation totals for 7 months during the middle of the growing season for maize (total growing season 10 months), 3 months during the middle of the growing season for potatoes at and below 3,050 meters, and 5 months during the middle of the growing season for potatoes in the higher elevations at and above 3,350 meters (7 months growing period). There are no potato fields in the Plaza farms between 3,050 and 3,350 meters. In accordance with the assumptions as to growing season, precipitation will be counted from December to June for maize, from August to October for potatoes which are planted and harvested in the same year, from November to January for low-elevation potatoes planted and harvested in different years, and from October to February for high elevation potatoes planted and harvested in different years. Precipitation data is taken from the station nearest the field in question. Most fields are nearest the Angla station; field T is nearest San Pablo, and fields Q, E, I, J, and S are nearest Zuleta. Certain missing months from the Angla precipitation sequence were supplied using averages between the San Pablo and Zuleta stations.

yielded up to 20,100 kilograms/hectare. For maize, the unfertilized plots produced about 1,100 kilograms/hectare; heavily fertilized plots produced up to 4,540 kilograms per hectare. A first impression is therefore that fertilization may more than triple the yields of maize and potatoes.

TABLE 2.6 PLAZA FARMS YIELDS CORRECTED FOR FERTILITY: MAIZE

Planting, maize	Nitrogen (kg./ha.)	Phosphate (kg./ha.)	Fertility index, I	Actual yield, Y	Corrected yield, Y/I
1	8	32	0.31	1.67	5.37
2	12	48	0.36	1.13	3.14
3	11	43	0.34	1.23	3.55
4	24	21	0.36	2.02	5.61
6	0	0	0.21	1.17	5.55
9	48	83	0.55	4.54	8.12
13	48	84	0.56	2.57	4.58
15	113	222	0.83	3.01	3.62
16	43	49	0.47	1.91	3.99
17	47	54	0.50	1.04	2.08

To analyze the role of soil fertility on yields, in Tables 2.6 and 2.7 the fertility index (I) for the various potato and maize plantings has been calculated. To help assure that initial soil quality is similar in the various plantings, fields which the record books indicate were planted to the same crop or to potatoes in the preceding year are eliminated from further consideration.[68]

In Tables 2.6 and 2.7 it is assumed that the initial soil phosphate is comparable to that in other relatively fertile highland locations (Table 2.1), or 38 kg/ha.[69]

Dividing yields by the fertility index gives an indication of expected yield under optimal fertilization. In general, this procedure gives reasonable results for crops produced under low fertilization levels.

Multiple regressions were performed relating potato yields and maize yields with the following factors: fertility index, precipitation, elevation,

68 Repeated cultivation of the same cultigen will presumably especially lower the nutrients which that cultigen demands. Since potatoes are normally heavily fertilized, crops which follow potatoes will benefit from residual fertility. Therefore we eliminate potato plantings 21, 22, and 24 (follow a previous potato planting by one year or less), maize plantings 8, 10, 11, and 14 (following a previous maize planting by one year or less), and maize planting 12 (a maize crop following a heavily fertilized potato crop lost to frost).

69 The average of the phosphate amounts in the sites tested by Bejarano, Lainez, and Portch, "Adequate Use."

TABLE 2.7 PLAZA FARMS YIELDS CORRECTED FOR FERTILITY: POTATOES

Planting, maize	Nitrogen (kg./ha.)	Phosphate (kg./ha.)	Fertility index, I	Actual yield, Y	Corrected yield, Y/I
1	21	85	0.46	1.5	3.22
2	0	0	0.21	5.9	27.99
3	31	125	0.57	13.8	24.36
4	18	71	0.43	9.1	21.22
5	19	138	0.52	10.0	19.22
6	141	191	0.81	10.7	13.15
7	65	102	0.63	8.0	12.74
12	46	102	0.59	14.1	24.06
13	91	193	0.79	14.3	18.20
14	92	208	0.81	2.7	3.37
15	53	127	0.64	17.6	27.42
16	86	201	0.79	16.1	20.44
18	67	170	0.73	13.4	18.44
20	75	139	0.70	14.9	21.25
26	110	282	0.87	11.6	13.29
29	46	137	0.64	11.5	18.08

and slope. Significant relationships with fertility and precipitation were indicated but the relationships with elevation and slope were not significant.[70]

PLAZA FARMS: ELEVATION AND PRECIPITATION

Simple linear regressions were performed between adjusted yield (Y/I) and precipitation, elevation, and slope. Surprisingly, slope was not significantly related to yield for either maize or potatoes.[71] Elevation was not significantly related to yield for potatoes, but was positively related to

[70] The StatView 512+ package was used for all statistical analyses. The beta coefficient table showed significance at the .05 level for potato fertility (.037) and precipitation (.047) but not elevation (.75) or slope (.52). Similarly, maize fertility (.016) and precipitation (.011) were significant at the .05 level but not elevation (.87) or slope (.17).

[71] Significant correlations were neither found for unadjusted yields nor for adjusted yields. Probabilities that correlations of slope and adjusted yield were due to chance were .6 for potatoes and .19 for maize.

yield for maize; for maize, yield increases with elevation.[72] For both potatoes and maize, precipitation had a significant negative correlation with yield.[73]

The lack of a negative relationship between yields and slope is particularly surprising in view of the fact that slopes are more likely to be eroded and less likely to receive supplemental irrigation in this area.[74] This evidence suggests however that Andean farmers maximizing short term benefits should farm irrespective of slope.

The consistent negative relationship between precipitation and yield is understandable in terms of the availability of supplemental irrigation in this area. The regression equations relating precipitation and yield can be used to calculate the hypothetical mid growing season rainfall amounts where expected yields decline to zero; these are 141 mm/month for maize and 191 mm/month for potatoes.[75]

In an effort to determine the upper and lower precipitation limits for potatoes, yields from plots with slopes greater than 12 per cent (and therefore inaccessible to supplemental irrigation) were plotted against monthly growing season precipitation and fitted with a polynomial curve. The result implied optimal yields at 95 millimeters per month rainfall.[76]

The lack of a negative relation between yield and elevation leaves unanswered the question of what determines the upper limit of cultivation for maize and potatoes. It suggests that efficient farmers should not have any bias toward lower elevations as long as labor inputs do not vary with elevation.

Plaza Farms: Crop Comparisons

It is useful to look at crop yields in terms of calorie production rather than in terms of kilograms per hectare. This not only gives an idea of the population which could be fed by the production of a given crop, but also allows comparison between crops. When corrected for soil fertility (by

---

72 Probabilities that correlations of elevation and adjusted yield were due to chance were .82 for potatoes but .03 for maize.

73 The r value for the correlation between adjusted potato yield and precipitation was -.769, and between maize yield and precipitation -.727. Both relationships are significant at the 2 percent level.

74 Slope is not significantly correlated with the residuals from the adjusted yield—precipitation regression for potatoes, and for maize the correlation with residuals is significant but in the opposite direction than expected—increased slopes seem to have higher yields.

75 The maize regression equation is $Y/I = -.083P + 11.74$ ($r = .727$), where P is mid growing season precipitation. For potatoes, the equation is $Y/I = -.201P + 38.48$ ($r = .769$). Yields, as always, are expressed as metric tons per hectare.

76 The fitted equation was $Y/I = -25.82 + .944P - .00524P^2$; $n = 13$, r squared is .612.

dividing by the fertility index) the calorie yields of potato and maize are not significantly different.[77]

The multiple linear regression of calorie yields of potato and maize fields combined against environmental factors produces the result that yields are significantly related to soil fertility and precipitation, and not to slope or elevation.[78]

If potatoes and maize are equivalent in productivity, there is at least one significant difference: maize requires a lower proportion of harvest for seed, especially at low production levels. The next chapter will show another advantage maize possesses which will help explain small farmer preference for maize over potatoes: a lower harvest labor requirement.

PLAZA FARMS: OTHER CROPS AND MULTIPLE CROPPING

The discussion so far has been limited to two crops, maize and potatoes, and has ignored intercropping.

A variety of other crops were planted in the Plaza farms, including barley and wheat. In a sample of 27, barley yielded about 20:1 (about 1800 kilograms per hectare) and wheat about 14:1; no relationship between yield and elevation could be demonstrated.[79]

Only two maize plantings are recorded as having been intercropped with beans. The yields of both were exceptionally high, even when adjusted for fertility and precipitation, although the sample size is so small as to be of no statistical significance.

Four maize crops and three potato crops were planted within a year's interval or less of a similar crop. Most of these successor crops were heavily fertilized (the one which was not, maize planting 14, was lost), so it is difficult to determine the effect of continuous planting on yields. Nevertheless, if one compares average yield of successor crops with the

---

77 Mean adjusted calorie yield per hectare of potatoes is 14.2 million (standard deviation 5.8) and of maize is 16.5 million (standard deviation 6.1). A small sample t test shows that the difference in means is not significant even at the 10 percent level. These "adjusted" yields are not the values most farmers experience with suboptimal fertilization.

78 Again, using the StatView package, and regressing calorie yield against the four x values: fertility (I), precipitation, elevation, and slope, only fertility and precipitation (t values 4.3 and 4.1 respectively) were significant at the 10 percent level.

79 A sample of 27 barley harvests was randomly selected from the account books which could be accurately located by name on the maps of agricultural fields. When a regression was performed of barley yield against field midpoint elevation, the correlation coefficient $r = 0.298$ was insignificant at the .05 level. A randomly selected group of 28 wheat plantings (1974-1979) also did not indicate a significant relation with elevation ($r = .085$).

average yield of the initial crops, and corrects for fertilization and precipitation, the successor crops only produced 62 per cent as much.[80]

Yield Constants: Yields at Median Precipitations

Observed maximum yields were 20.1 metric tons of potatoes per hectare, 4.5 metric tons of maize , and 3.7 metric tons of barley. When corrected for fertilization (Y/I), maxima were even higher: 28.0 for potatoes and 8.1 for maize. Such yields obviously cannot be expected year after year because of variations in climate. The "yield constant" or median corrected yield at this location can be determined by calculating the expected yield at the median precipitation at the Plaza farms. Such a procedure gives the values of 3.98 tons/hectare maize and 19.62 tons/hectare potatoes.[81]

The yield constants give the expected yields during the median years at this site. If fertilization is known, the yield for a "typical" year at this site can best be calculated by multiplying the fertility index by the yield constant. In places like the Plaza farms, where yields decrease above the typical precipitation level and increase below the typical precipitation level, the yield constant provides an acceptable estimate of average yields as well. If, however, yields decline both above and below the median precipitation value, as on the dry margin of cultivation without irrigation, the yield constant gives an "optimum" yield value for the site and should be adjusted downward by an adjustment factor to give a better estimate of median yields.[82]

---

80 For maize, referring to Table 2.5, Plantings 14 and 8 followed plantings 12 and 6 respectively; planting 11 followed planting 10 which followed planting 9. For potatoes, plantings 21 and 22 occupied parts of the field occupied by planting 20; planting 24 followed 21 and 22. Yields corrected for precipitation and fertility (doubly corrected yields) were calculated by dividing observed yields by the fertility index I multiplied by the factor (-.201P + 38.48) for potatoes, or (-.083P + 11.74) for maize; these latter are based on the regressions of corrected yield against precipitation discussed in a preceding footnote. The doubly corrected yields should be unity if all else is equal. In the event, the average doubly corrected yields for the initial plantings (20 of potatoes, 12, 6, 9 of maize) was 1.42 (perhaps accounting for the risk of planting an immediate successor crop), while the average doubly corrected yields of all the successor potato and maize crops was 0.88.

81 Using Tables 2.6 and 2.7, median precipitation at the Plaza Farms is 94 millimeters per month. The yield constants are calculated using regression equations between precipitation and corrected yield, using the data of Tables 2.6 and 2.7.

82 At the Plaza farms, the yield constants are in fact close to median and average yields.

The Production Investigation Program Data

Although most agricultural research station activity is not directed toward the understanding of small farm agriculture, in the northern highlands of Ecuador a very useful program has been the Programa de Investigación en Producción (PIP), devoted to the study and improvement of small farms.[83]

TABLE 2.8 PIP DATA ON MAIZE YIELDS

Planting	Nitrogen added (kg./ha.)	Phosphate total, soil plus fertilizer (kg./ha.)	Fertility index	Yield (metric tons/ha.)
Quinchuqui 1	30	166	0.61	1.94
Quinchuqui 2	59	138	0.67	1.6
Quinchuqui 3	59	166	0.71	2.05
Quinchuqui 4	59	190	0.73	1.66
Quinchuqui 5	73	178	0.75	2.64
Quinchuqui 6	87	166	0.75	2.01
Quinchuqui 7	87	190	0.78	2.63
Quinchuqui 8	87	214	0.8	2.64
Quinchuqui 9	115	190	0.8	2.72
Imbabuela 1	30	102	0.53	2.1
Imbabuela 2	59	74	0.57	1.7
Imbabuela 3	59	102	0.62	2.2
Imbabuela 4	59	126	0.66	2
Imbabuela 5	73	114	0.66	2.2
Imbabuela 6	87	102	0.66	2.7
Imbabuela 7	87	126	0.7	2.6
Imbabuela 8	87	150	0.73	3
Imbabuela 9	115	126	0.72	3

Source: Instituto Nacional de Investigaciones Agropecuarias (INIAP), *Programa de Investigación en Producción: PIP Imbabura: Informe Annual 1983*, Quito, Ecuador, 1983.

83 See INIAP (Instituto Nacional de Investigaciones Agropecuarias), *PIP Imbabura: Informe Annual 1983, Programa de Investigación en Producción*, Quito, 1983; and INIAP (Instituto Nacional de Investigaciones Agropecuarias), *PIP Imbabura: Ciclo Agrícola 1984-1985, Programa de Investigación en Producción*, Quito, no date. Data in this section are taken from these two reports. Other reports have also been consulted, including those concerning a project in Cayambe.

This program has been active in exploring the utility of new varieties of maize and fertilizers for traditional small farmers. In the 1982–83 trials, two sites in Otavalo province (near the Plaza Farms) were among the sites tested and produced yield results for various fertilizer trials (Table 2.8). These sites, Quinchuquí cooperative and the farm of Andrés Maldonado at Imbabuela, both near 2,650 meters elevation showed clear responses of maize yields to fertilization in a manner consistent with the results from Zuleta.[84] Maize yields from these sites are not notably higher than those from the higher fields at Zuleta, even when corrected for fertility.[85]

The Best Land

The results above can be summarized as indicating that the best land for contemporary Ecuadorian large highland farmers is not necessarily limited to lower elevations or to gentler slopes. Yields (in calorie terms) are constant or even increase up to the upper limit of cultivation and on steeper slopes. Farmers therefore should not be expected to avoid steep slopes or higher elevations on principles of yield response, although the comparability of potato and maize agriculture with respect to labor efficiency still needs to be explored (see below). Precipitation is a major limiting factor, with excess precipitation being perhaps as significant as deficient precipitation. The lower limit of rainfed agriculture, as determined from actual distribution, may approximate 400 millimeters for barley and 600 millimeters for maize; the upper limit appears to be about 1,400 millimeters, consistent with data from the Plaza farms. Soil fertility is exceptionally important as well for yields; even under fertilization, yields decline under continuing cultivation.

There are no small-farm data in Ecuador of sufficient quality to test these ideas in the same way. In the next chapter, other sources of information on small farm adaptive strategies will be analyzed.

---

84 At Imbabuela, the yields are related to the fertility index with an r squared of .657; and at Quinchuquí, with an r squared of .509 (both significant at the 5 percent level). Both sites were planted to maize variety "INIAP-101." Trials with traditional maize showed yields which were substantially similar. At a third trial site in Otavalo cantón, Pijal Bajo, yields were high even for minimal fertilization (in the "fertilización niveles" experiment) so no relation with fertility index could be demonstrated; curiously, much lower yields were found in the "verificación de tecnologías' experiment, consistent with the other two sites. The PIP trials have been plagued with problems of less than perfect farmer cooperation.

85 At Zuleta, the regression of yield against fertility index suggests an optimal yield of 4.2 metric tons per hectare with a fertility index at 1; at Quinchuquí, the corresponding value would be 3.6 metric tons, and at Imbabuela, 4.3 metric tons.

Chapter 3

SMALL-FARMER ADAPTIVE STRATEGIES

The Farm Family

The small subsistence farmer has been the numerically dominant element of the rural equatorial Andes. In 1974, 75 percent of the farm families in the equatorial Andes owned less than 5 hectares; these small farms produced 63 percent of the total maize harvest, despite their control over only 11 percent of the land.[1]

The small farms of the highlands primarily produce subsistence crops for the families themselves. Such production for the needs of the household is subsistence production in the strict sense of the term.[2]

We cannot, however, speak of a subsistence *economy* in the highlands today because production for subsistence is always combined with various tactics to earn cash. This often takes the form of migration of family members to nearby cities (especially Quito and Guayaquil) for wage labor. Combinations of subsistence production and migration or commodity production tend to be persistent because the subsistence activities ensure the producer's low cost of living and child raising, and enable the producer to sell goods or labor below the acceptable price of capitalist producers or full-time wage laborers who must support families with purchased goods.

---

1 See PRONAREG-ORSTOM (Programa Nacional de Regionalización Agraria – Office de la Recherche Scientifique et Technique Outre-Mer), *Documentos 1–11, Diagnóstico Socio-Económico del Medio Rural Ecuatoriano*, Quito: Ministerio de Agricultura - ORSTOM, 1978, Volume A, 86, and Document 4A.

2 Small farms (0–5 hectares in size) in Tabacundo, near Cayambe, use rather than sell 95 percent of the maize they produce; see MAG-INERHI (Ministerio de Agricultura - Instituto Ecuatoriano de Recursos Hidraulicos), *Proyecto Tabacundo, Tomas de Estudio de Factibilidad*, Quito (7 volumes), 1978. For a good discussion of subsistence production see Georg Elwert and Diana Wong, "Subsistence Production and Commodity Production in the Third World," *Review (Binghamton, New York)*, 3 (1980), 501–522.

Pre-capitalist production relations have also persisted in the highlands with the so-called "traditional" haciendas. These large land holdings have in the past elicited unpaid labor in exchange for *huasipungos*, or in exchange for such services as access to grazing land, wood, or use of hacienda roads. Even today vestiges of these arrangements persist in remote areas.[3]

In recent years, many haciendas (including the Plaza farms) have made the transition to "modern" capitalist farms; these farms concentrate on the production of commodities (potatoes, dairy products, and grain) for urban markets, using wage-earning workers and occasionally capital-intensive techniques. In general, however, the haciendas of highland Ecuador now play a very limited role in providing employment or services to the rural population, so that it is somewhat misleading to suggest that there is a socio-economic "structure" dominated by large farms. Large farms are perhaps most significant in continuing to deny small farmers access to certain locational niches, particularly the flats.

Vegetation in Cultural Context

The higher elevations, above about 3,000 meters, are almost entirely in bunch grass. Farmers agree that this is due to burning. Brush is not valued highly because (1) brush land harbors birds which eat seeds and damage crops; (2) brush is more difficult to clear for crops than grass (see below); (3) grass is considered to produce a valuable fertilizing ash when burned; (4) grass is a valuable resource for grazing sheep and cattle; (5) grass can be used as fuel for household cooking;[4] and (6) children enjoy setting fires while tending animals in the páramo.

On slopes above 3,000 meters south of Colta, it was reported that páramo grass always had to be burned to prevent shrub invasion. In abandoned fields, weeds and small shrubs invade; these are burned or turned under. Moist flats are also invaded by brush; the flats at Zuleta have been cleared of brush in recent memory. At Zuleta, above 2,800 meters, it is still possible to see *jura p'anga* (Q.) (*Solanum* sp., a woody shrub) pioneering the invasion of the flats.

---

3 Huasipungos were small plots given in exchange for labor services. See PRONAREG-ORSTOM, *Diagnóstico, Documentos A–D*; especially Document A; CIDA (Comité Interamericano de Desarollo Agricola), *Ecuador, Tenencia de la Tierra y Desarollo Socio-Económico del Sector Agrícola*, Washington D.C.: Panamerican Union, Organization of American States, 1965; Alfredo Costales and Piedad Peñaherrera de Costales, *Historia Social del Ecuador, Llajta 17–19*, Quito: Instituto Ecuatoriano de Geografía e Historia, 1964; Osvaldo Barsky and Gustavo Cosse, *Tecnología y Cambio Social: Las Haciendas Lecheras del Ecuador*, Quito: Facultad Latinoamericana de Ciencias Sociales, 1981.

4 Weismantel, *Food*.

Below about 3,200 meters, burning does not seem to favor the generation of bunch grasses. Farmers express fear of the propagation of kikuyu (*Pennisetum clandestinum*), which is difficult to remove.[5] Often disturbance leads to the dominance of a variety of exotic weeds and shrubs, in extreme cases favoring wind erosion.

Most of the economically important non-cultivated plants are not located in brush or woodland (*monte*) but rather in the hedgerows which are such a conspicuous feature of much of the highland landscape. Hedgerows often are a complex environment incorporating walls, embankments, drainage ditches, or canals. A list of some of the more characteristic plants of hedgerow and fallow field may be found in Table 3.1.[6] Other plants of hedgerows include solanaceous fruits (*chímbolo*, *uvilla*), and amaryllids (*Agave*, *Fourcroya*), which can be a source of an unfermented beverage. Another edible plant is wild mustard (*nabo*). Mushrooms (*kallampa* Q.) are eaten. There is little evidence, however, that the volunteer plant resources of hedgerow and field play a fundamental role in the calorie supply of the diet or that they would be long suitable as famine foods.

Animals in Cultural Context

The wild fauna of the highlands is now rather impoverished. Perhaps the most important large animal from the peasant viewpoint is the fox (called paradoxically not *zorro* but *lobo*, *Canis Azarae*). In some cases, as in the flats and slopes near San Pablo, foxes may be sufficiently numerous as to render impossible the keeping of household animals such as chickens, even at times killing guinea pigs in houses.

Deer are present, but do not contribute to the diet. Bear are rare. Rabbits (conejo, *Lepus* sp.) are sometimes eaten. Among the birds, gallinaceous fowl or "partridges" (*perdices*, *yutu* Q., *Attagus* spp. and *Crypturus* spp.) are attractive for their meat, as are doves (*tórtola*, *urpi* Q., *Columba passerina*).[7] Landowners remember that migratory ducks were

---

5 The difficulty of removing kikuyu can be a factor discouraging long fallows, although the grass, by providing productive forage, aids in cattle grazing and manure production for existing fields. Farmers now may see field abandonment less as a means to regenerate in situ soil fertility than as a means to raise cattle and indirectly increase manure production for existing fields.

6 This table only contains personal observations and reports from informants; it could have been extended and completed using references from the literature. Plants were identified primarily by Henrik Balslev of the herbarium of the Pontifical Catholic University of Ecuador; I was also helped by Hugh Iltis of the Herbarium of the University of Wisconsin at Madison.

7 See Wolf, *Geografía*, 506.

TABLE 3.1 HEDGEROW AND FIELD PLANTS, SAN PABLO, ECUADOR

Local name	Scientific name	Situation	Uses
Pima	Calceolaria tripartita R.&P.	Ditches	Eaten with chocho
Ichu Jiwa	Polygonum persicarioides H.B.K.	Ditches	Not eaten
Izu	Dalea cylindrica Hook	Ditch sides, embankments, etc.	Cough medicine
Jura p'angu	Solanum sp.	Open sunny areas; pioneer	Firewood; leaves used in chicha making
?	Monnina sp.	Shady banks above drainage ditches	
Alpa Quinua	Chenopodium sp.	Sunny fallow field	Bitter as food (informant)
Bledo	Amaranthus quitensis HBK	Agricultural field weed	Eaten young
Malva	Unidentifiable	Fallow fields	Forage
Yerba de Cuy	Compositae	Fallow fields	Guinea pig food
Silantro	cf. Coriandrum sativum	Dense stands in fallow fields	Condiment
Pacunga	Bidens pilosa L.	Fallow fields	—
Forasterita	Silene sp.	Fallow fields	Guinea pig food
Ambo	Unidentified	Fallow fields	Forage
?	Castilleja fissifolia Sesse & Moc.	Isolated plants in sunny disturbed areas	—
Mora	Rubus floribundus H.B.K.	Common on hedgerow banks	Edible berries
Lechero	Euphorbia latescens	Common in fences, hillside shrub or dooryard tree	Used as "living fence"

TABLE 3.1 *CONT*. HEDGEROW AND FIELD PLANTS, SAN PABLO, ECUADOR

Local name	Scientific name	Situation	Uses
Chilca	Baccharis sp.	Common in hedgerows, roadsides, slopes	—
?	Holcus lanatus L.	Grass on bank tops	—
Zoresno	Cassia sp.	Banktop shrub and dooryard tree	—
?	Miconia papillosa (Desr.) Naud.	Hedgerow banktop shrub	—
?	Bidens humilis H.B.K.	Common in sunny pastures, slopes	—
Avena	Avena sativa L.	As common weed in maize fields	—
Yerba mora	Solanum americanum Mill.	Common on hedgerow banks	—
?	Alternanthera sp.	Common shrub on hedgerow banks	—
?	Oxalis sp.	Hedgerow banks	—
?	Calceolaria crenata Lam.	Side of ditches	—
?	Gomphrena sp.	Side of ditches	—
?	Salvia sp.	Side of ditches	—
Higuerilla	Ricinus communis L.	Common in hedgerows	Oil used as varnish
Marka	Not identified	Hedgerows	Insecticide for fleas; medicinal for hemorrhoids
?	Datura stramonium L.	Sunny fallow fields	—
?	cf. Cestrum	Ditchside shrub	

once numerous in humid flats such as those around San Pablo Lake; they are now rare due to drainage of flats and excessive hunting. Smaller birds are still hunted by small boys with slingshots in brushy areas.

The only native highland fish is the *preñadilla* or *chalwa* (Q.) (*Pimelodus cyclopum*).[8] Other fish such as trout have been introduced. Fishing is officially prohibited in San Pablo Lake, but is still practiced at Colta Lake.

Perhaps the most important fauna from the dietary point of view is the snail (*churu* Q.) and the June bug (*katsu* Q.) and its grub (*juzu*). The latter may be cooked with potatoes or rice, or eaten fresh when turned up by the plow; the adult June bug is eaten, toasted and with the legs removed, about November.

Tools

Perhaps the most important hand tool is the *azadón* (*asadun* in Quichua), a kind of short hoe which is suitable for working the silty highland soils, weeding, making ridges and furrows, and harvesting potatoes. Near Colta, the handle is made from eucalyptus wood, while the steel blade is purchased for about $9 U.S. (1981).

The *barra* is a long heavy rod of metal with a sharp wedge-like tip; it is used to hack out blocks of sod and in heavy excavation chores such as ditch construction. The *lampa* and *pala* are kinds of spade; they may be used to remove loosened earth from a ditch in the process of construction, or to remove muck from ditches during cleaning. The *pico* is a pick which is used for such tasks as breaking up soil hardpan. At times a long stick (*palondra, chaquín*) is used for planting maize.[9]

Harvesting of potatoes and other tubers involves use of the azadón, which flips the tubers out of the ground. Harvesting of grains often requires use of the *hoz*, a sickle.

Crops are put in sacks (*costales*) and carried out of the field on human backs or on animals.

Land Reclamation

Two alternative practices may be followed in reclaiming land from brush: slashing and incorporating the rotting material into the soil (*pudrición*) and slash and burn (*quema*).

1. *PUDRICION*. In this technique, the brush is cut; a second cutting (*quitar retoño*) may also be required. If the brush is tall, the branches are then cut into smaller pieces (*repicar*). The material is

---

8 Wolf, *Geografía*, 506.

9 See Costales and Costales, *Katekil*, 224–225.

then left to rot for one to two years. Finally the material is turned over (*tolar)*.

Based on several field interviews, the amount of labor time involved in pudrición (including tolar) is 60 to 80 person-days per hectare, relying mostly on male labor with machete and pico.[10]

2. QUEMA. This process is similar to pudrición, but instead of waiting for 1 or 2 years for the stems and roots to rot, the brush material is immediately burned and the roots are grubbed out. Because of the burning, much of the potential contribution to soil fertility is lost.[11] One source suggested that 108 person-days per hectare are required to remove brush and roots.

In reclaiming land from páramo grass, there are also several possibilities.

1. POTATOES MAY BE SOWN UNDER SLITS CUT INTO PARAMO SOD. This is said to require little labor and to yield well, although it is uncommon and was not observed.
2. *VOLTEAR TIERRA*. In this process, sod is pried up and turned over to rot for a few months. With modern metal tools, from 20 to 54 person-days may be needed for this step per hectare. With ox plow, only 6 person-days per hectare may be needed.
3. BURNING OF PARAMO SOD. Instead of turning under the sod to rot, the sod may be piled up and burned. The ash is then spread over the fields as fertilizer. As practiced in Chimborazo, 48 person-days are said to be needed to remove the sod blocks (*tsagmana* Q.), 12 more person-days/hectare to pile and burn the blocks, and another 12 person-days to spread the ash around, for a total of 72 person-days per hectare.
4. HERBICIDE. Herbicide has been used on the Cajabamba flats near Colta to remove kikuyu grass. The cost was $250/hectare (US);[12] this amount of cash would have bought 150 person-days of labor,

---

10 In estimating the labor inputs for this and other agricultural processes, it is worth bearing in mind that (1) the time involved varies between men, women, and children; (2) the intensity of labor and time worked per day vary from family to family; (3) family labor may involve a different amount of time than paid (*jornal*) labor; (4) tasks may involve work parties of neighbors who are paid in kind, or a number of different family members; and (5) a considerable amount of time may be involved in simply travelling to and from the work site. Here it will be assumed that "person-days" of labor do not include travel time, involve the combination of ages and sexes most typically involved in the task, and consist of 8 hours. Real situations may involve wide variations from the figures given.

11 Quema may also be used to clear land of crop remains prior to sowing. A farmer near San Pablo told me that this provided more needed "heat" for the soil.

12 Philip Westra, personal communication.

and very likely is less attractive where labor is available for more traditional methods.

In areas reduced to *mogote* (exposed hardpan), reclamation involves not so much destruction of vegetation as breaking up of soil and adding fertilizer. According to an informant near Colta, breaking up the hardpan with a pick requires 100 person-days per hectare. The blocks would then be allowed to soften in the rain, be pulverized (*ñutuchina* Q., *pedazear*), and be fertilized with 330 *costales* (3,300 kilograms dry weight equivalent) of manure per hectare.

Table 3.2 summarizes the above discussion, giving alternative labor inputs for different reclamation stages. For the purposes of developing the model used in the next chapter, it is useful to relate reclamation labor inputs to fallow length. The following equation, relating reclamation labor (L_r) in person-days per hectare to fallow time (r) in years is consistent with the data presented in Table 3.2:

$$L_r = 20.1r^{0.28}$$

TABLE 3.2 LABOR INPUTS, MAJOR ANDEAN CROPS

	Person-days per hectare		
STEP	POTATOES	MAIZE	BARLEY
Preparation:			
Brush/Quema	110	110	110
Brush/Pudrición	70	70	70
Grass/Quema	70	70	70
Grass/Pudrición	37	37	37
Herbs	20	20	20
Unspecified	27	16	19
Sowing	11	10	7
Cultivation	60	30	10
Harvest	20–60 (10 plus 3/ton)	10	17

Pests

Traditional farmers find fallow and rotation management especially useful in pest management, but in general there are few traditional tactics to control pest attacks per se.

Farmers frequently use fungicides and other pesticides, especially on potatoes and irrigated crops, where sale helps repay the purchase costs. Pesticide use is often dependent on rainfall, with low use during dry periods and high use during wet periods when infestations are greater.

Large farms can afford to apply fungicide and insecticide as a matter of course to each crop.

As of early 1987 free importation and marketing of pesticides was permitted, and there was no control of pesticide use; the result is widespread abuse and excessive application of chemicals, including chlorates banned by other countries. Contamination of milk and meat has been found.[13]

Fertility and Fertilization

The importance of animal manure in indigenous farming has tended to be underestimated.[14] Part of the problem may be that the words for fertilization (*abonear*) and fertilizer (*abono*) in Spanish are usually understood to refer to chemical fertilizer only. The word for manure is *majada* (*wanu* Q.); the former word also is used to refer to dungheaps in the houseyard. Even the Spanish word for manuring (*estercolar*) may be limited to the practice of consciously manuring each plant, and not include the use of movable corrals ("folding"), throwing manure into fields (*botar*), or incorporating organic refuse into the soil (*tolar*). Practically all highland farmers in all regions appear to use animal manure in fertilizing their crops.[15]

The need for such fertilization can be appreciated by examining the nitrogen and phosphorus removed by crops. For even the low 617 kilograms/hectare yields typical of the Ecuadorian highlands, 16 kg/ha of nitrogen and 6 kg/ha of phosphate are removed from the soil.[16] Since typical manures are only 2 or 3 percent nitrogen by dry weight, it can be appreciated that a substantial amount of fertilization would be required to replace these nutrients. The quantity and quality of this manure, as well as fertilization practices, will be discussed in the remainder of this section.

Most animals produce about 4 kilograms of manure a year (dry weight) per kilogram of live body weight.[17] The various animals, including human

---

13 USDA (United States Department of Agriculture), *Ecuador, Annual Agricultural Situation Report*, Quito (Mimeographed), 1987, 21.

14 Except by Bruce Winterhalder, Robert Larsen and R. Brooke Thomas, "Dung as an Essential Resource in a Highland Peruvian Community," *Human Ecology*, 2 (1974), 89–104. Also see Treacy, *Fields*, 284–286.

15 Based on personal observations and interviews. At Chauzán and Galte, farmers were explicitly asked about animal fertilizer use; all replied in the affirmative. See Cañadas, Salcedo, Bravo and Knapp, "Vulnerability," 419.

16 Roy Donohue, J. Schickluna and L. Robertson, *Soils: An Introduction to Soils and Plant Growth*, Englewood Cliffs, New Jersey: Prentice-Hall, 1971, 320; values prorated for 617 kilograms/hectare.

17 Nyle C. Brady, *Nature and Properties of Soils*, New York: Macmillan, 1974, 535.

beings, present in the farm household constitute one of the major sources of fertilizing material.

An adult of 130 pounds could be expected to produce 227 kilograms per year of night soil. A household of five very conservatively must produce at least 700 kilograms per year.

Guinea pigs are present in most indigenous households. A household in San Bernardo, Chimborazo, chosen for a guinea pig manure production test produced 15 to 29 kilograms a year of guinea pig manure per guinea pig (dry weight equivalent); several interviews tended to corroborate these figures.[18]

Measurements of sheep manure production while corralled at night indicate that they produce about 70 kilograms/year of manure each. This very conservatively implies 112 kilograms/year each of manure, including that produced during the day while away from the corrals.[19]

Manure production from cattle, pigs and chickens was not directly measured; farmers in Chimborazo estimated cattle manure production at about 450 kilograms/year, pig manure production at about 55 kilograms/year, and chicken manure production at 10 kilograms per year. These values are considerably lower than United States values.[20]

---

18 The household had 10 guinea pigs fed with grass, meadow weeds, and household scraps. The household floor was first cleaned normally and 17.3 kilograms of moist feces (with a small amount of intermixed grass) was produced, said to have been the result of 9 days' accumulation. When the housewife saw the general interest in the experiment, she cleaned out 7.3 kilograms more, probably not produced within the last 9 days. After 5 days I returned and the wife collected 5.0 kilograms more of moist guinea pig manure. The manure was tested at the Polytechnic School of Chimborazo and determined to have a moisture content of 59 percent. Thus, the above measurements suggest that from 15 to 29 kilograms/year of guinea pig manure (dry weight equivalent) are produced per animal per year.

19 Manure was collected from a corral where 6 sheep (2 large and 4 small) were penned for 15 nights (from about 4:30 pm to 7:30 am). Total production was 17.3 kilograms (dry weight), which corresponds to 70 kilograms/sheep/year. This represents manure production during only 15 hours of the day; if production during the rest of the day is similar, these sheep produce 112 kilograms/year each of manure. Actually, this figure is conservative, since animals produce most of their feces during the day rather than at night (see J. Azevedo and P. R. Stout, *Farm and Animal Manures: An Overview of Their Role in the Agricultural Environment*, Berkeley: California Agricultural Experiment Station Manual 44, 1974). The only previous report of sheep manure production from highland Ecuador suggested 197 kilograms/year, without indicating the source for this figure: see CIDA, *Tenencia*, 254.

20 All values are converted to dry weight equivalents. See Azevedo and Stout, *Farm and Animal Manure*, 8–12, for estimates of manure production in the United States.

Table 3.3 lists the quantities of manure produced in a typical 1–2 hectare highland farm, based on average numbers of animals[21] and the manure production measured or reported in Ecuador.[22] The same table indicates estimates of nitrogen and phosphorus production for each manure category.[23]

TABLE 3.3 MANURE PRODUCTION IN A TYPICAL HIGHLAND FARM

Animal	Quantity	Manure/year (kg, dry)		N (kg.)	P2O5 (kg.)
		PER ANIMAL	TOTAL		
Human	5	140	700	20	9.9
Guinea pig	20	22	440	13	6.2
Sheep	4	115	460	15	4.1
Cattle	1	450	450	10	4.1
Pigs	3	55	165	5	2.3
Chickens	7	10	70	2	2.1
TOTAL			2,285	65	29.0

Only about 40 percent of the nitrogen and 60 percent of the phosphorus in manure may be available to plants.[24] Thus, using the data in Table 3.3, about 26 kilograms of available nitrogen and 17 kilograms of available phosphate per year may be available to a typical highland family.

Manure is collected in a variety of ways. Guinea pig manure and house trash is scooped up by hand by the housewife, once or twice a week;

21 Estimates of 5 persons and 20 guinea pigs on a typical farm are my own; the other values are taken from PRONAREG-ORSTOM, *Diagnóstico, Documentos 1–11*, Document 5, 32.

22 Other modern dung sources include burros, horses, oxen, and goats; dogs also provide some manure.

23 The values given for sheep, cattle, pigs and chickens are 80 percent of the average of the values given for the United States in Azevedo and Stout, *Farm and Animal Manures*. A lower than American value is suggested by the results from Nuñoa, Peru, where values about 60 percent of US means were obtained; the Nuñoa samples were old, however, having been sent to the United States for analysis. See Winterhalder, Larsen, Thomas, "Dung." The values for guinea pigs and humans are based on the average of values for sheep, cattle, pigs, and chickens.

24 Rule of thumb used by the University of Wisconsin Soil Plant Analysis Laboratory. However, as manuring continues, further nutrients are released into the soil, and the organic matter in manure seems to increase the availability of phosphorus. See Azevedo and Stout, *Farm and Animal Manures*, 67 and 71; and Louis Thompson and Frederic Troeh, *Soil and Soil Fertility*, New York: McGraw Hill, 1973, p. 267.

this is carried out and thrown onto the fields. Little piles of house trash mixed with ash and guinea pig dung are a common sight in Chimborazo fields.[25] Straw from house floors may be quite high in nitrogen because of the presence of guinea pig urine.[26]

Farmers defecate and urinate in fields near their houses. The manure is considered rather too strong, but still valuable. It is apparently not stored, and latrines are traditionally not used;[27] therefore there is no loss of nutrients due to storage.

Sheep spend the night in corrals; during the day they are led out to graze in harvested fields, fallows, pastures in the flats, or the páramo grasslands. The night corral may be next to the house, in which case the dung is periodically swept up with a broom of branches and leaves (*p'ichana* Q.), either to be brought out to the fields in a sack (*costal*) on burro back, or to be stored on the dung heap. Alternatively, sheep may spend the night in movable corrals (*redil portatil, quincha móvil, talanquera*). These corrals are 5 to 40 square meters in size, and are moved every three nights or so; in the course of a year a considerable fraction of a hectare can be heavily fertilized [28] The corrals are constructed of light wooden poles, which can be readily set up even on fairly steep slopes; flat surfaces or terracing are not required.

Pigs are kept in an enclosure near the house, often a cave dug into a semi-terrace (*mojón*) face. This manure is collected and used similarly to sheep manure. Cattle manure is considered inferior to sheep manure, but is used; the family cow may be tied up at night near the house and its manure periodically collected. Similarly, burro dung is collected and used.

Manure is normally not collected from distant pastures. Farmers in grassy flats will, however, collect dung from neighboring pastures to fertilize their fields.

Chickens must be kept in a location safe from predators, such as a tree. Their manure is collected and is highly prized, as it is rich in phosphorus.

---

25 The use of guinea pig dung as a fertilizer was also reported for Imbabura over 25 years ago; see Rubio Orbe, *Punyaro*, 34. Occasionally, house trash is not used but simply dumped on waste land or in a gully.

26 A sample of house floor cleanings from near Lake Colta was analyzed as 1.3 percent nitrogen by weight.

27 Galo Plaza Lasso told me the story that his installation of latrines in the area of former huasipungos near Zuleta was resisted by farmers on the argument that it would be a sin to waste such a god given resource as night soil.

28 A survey of the Chauzán and Galte communities in Chimborazo showed that 47 percent of the sample used movable corrals. Users of the corrals had at least 20 animals, with a median herd of about 30; the typical corral was 20 square meters and was moved every three nights. If each animal produced 0.2 kilograms manure per night, these corrals deposit about 0.9 kilograms of manure per square meter, or 9000 kilograms per hectare: perhaps 293 kg/ha of nitrogen and 80 kg/ha of phosphate. See Cañadas, Salcedo, Bravo, and Knapp, "Vulnerability," 419–421. Also see CIDA, *Tenencia*, 251, and Costales and Costales, *Katekil*, 60.

The most intensive technique of fertilizing is the *shuglutsi* (*puñado*) method, practiced both near San Pablo and Lake Colta. At the time of potato sowing, one person makes a furrow with the azadón, while another plants potatoes and inserts a fistful of dung (shuglutsi). The fistful is about 75 grams dry weight equivalent; this application rate implies about 1,500 kilograms of manure is applied per hectare. There is often a second application of manure.[29]

Maize fields are generally fertilized broadcast with manure. During the dry summer season, it is common to see large manure piles in maize fields near San Pablo ready to be spread prior to sowing. Fertilizer applications are lower for maize and other grains than for potatoes.[30]

A source of fertility apart from human technology is the droppings from earthworms and juzu grubs. I collected juzu dung from an old raised field soil in Imbabura; the dung had 86 times the nitrogen concentration of the surrounding soil.[31] Farmer preference for *yanallpa* (black soil) rich in *kuika* (earthworm casts) is logical due to the fertility-renewing activities of soil fauna.

The practice of *tolar* (incorporating organic matter in the soil prior to sowing) may involve the incorporation of crop remains (maize stalks, field weeds, or even Eucalyptus leaves). At the same time, any manure, house trash, or ash on the field is also incorporated in the soil.

As seen in the preceding chapter, the use of chemical fertilizer is not unknown in Ecuador. It is most commonly applied to potatoes, even by small farmers, although the latter have difficulty in getting soil samples and technical advice to apply the appropriate concentrations of the appropriate nutrients. Most of the fertilizers used in the highlands are formulated fertilizers, and are mostly produced by FERTISA, a company with majority state ownership and relatively low profits.[32]

---

29 A sample "fistful" was collected near Colta from a farmer and oven dried and weighed. The rate per hectare is based on typical potato spacing (50 cm in one direction, 90 cm in the other).

30 One informant suggested an application of 1,800 kg/ha manure to achieve a yield of 1,400 kg/ha of maize.

31 The dung sample and soil sample were both tested at the Escuela Politécnica de Chimborazo in Riobamba. In the United States, earthworm droppings are also rich in nitrogen and phosphorus, and may amount to as much as 18,000 kilograms per hectare. See Brady, *Nature and Properties*, 117.

32 The United States and West Germany have been the chief suppliers of fertilizers to FERTISA; variations in fertilizer sales have been great from year to year due to weather problems and pricing fluctuations. See USDA, *Ecuador Situation Report 1987*, 20–21 and Tables 4.3 and 4.4.

Semi-Terracing and Terracing

During the process of cultivation, notches are made at the edges of fields. Notches are deliberately made to indicate property boundaries and to limit cattle movements. Property boundary notches are found at the edge of the linear fields of the Colta flats; these resemble (but are not) drainage ditches. The most common term for such marked field boundaries is *mojón* (plural, *mojones*). On slopes, the notches or nicks on the upper sides of fields grow over time, becoming a riser or wall separating two fields. This riser too is called a mojón. The major functions continue to be to delimit property and limit animal movement.

The risers often appear to be terrace walls, but farmers always insist that they are not terraces; indeed, there appears to be no indigenous word for terrace in the Ecuadorian agricultural lexicon. *Andén* is not used. The semi-terrace risers are called *mojones* or at times *mogote* (exposed hardpan).

With continuing cultivation the risers may become two or more meters high. The silt soils easily maintain vertical faces, although larger faces may be reinforced with blocks of hardpan to prevent them from falling.

The top of the riser may be bare or it may have a line of hedgerow shrubs (Table 3.1), cactus, eucalyptus trees, or grasses (*fihuíl*). These plants may serve to trap some material eroding from above.

Normally the mojones are not true terraces, since there is not much reduction of slope or accumulation of soil; on the contrary, they are erosional features which signal removal of a large percentage of the original topsoil layer. Under certain circumstances, however, the mojones do serve some functions. On very steep slopes, closely spaced mojones certainly facilitate hand-tool farming by reducing slope, even though they represent erosion of topsoil.

True terracing is occasionally present in the equatorial Andes. In eroded areas with large amounts of exposed hardpan (mogote), blocks of hardpan (*molones*) may be piled into walls to accumulate eroding soil.[33] An informant estimated labor inputs for wall construction with molones. The mining process requires about 5 minutes per molón, for a man working with a pick. A wall one meter high requires 1 person-day of mining labor for every two horizontal meters. In addition, portage requirements are 20 round trips per meter of wall. It is easy to show that portage labor is greater than mining labor if the blocks are mined over 300 meters from the wall.[34] The parts of the equatorial Andes which lack hardpan or accessible stone tend to lack terraces in part because transport is prohibitively expensive. As will be seen below, the volcanic ash soils permit other strategies than

---

33 Such walls are common near Punín, Chimborazo, and are currently being constructed on slopes to the east of Lake Colta.

34 Assuming a speed of 24 kilometers per day, 20 round trips would require a half a day if the source were $(12 \div 2) \div 20 = 0.3$ kilometers away.

terracing to permit irrigation; it has already been demonstrated that fertilization can take place without terracing.

Indeed, there are very few modern or prehistoric agricultural terrace walls in the equatorial Andes (see Chapter 5).[35]

Drainage

On the slopes, cultivation and the creation of semi-terraces normally result in the development of a system of drainage ways. Sometimes these drainage ways are gullies which have developed through natural processes, although usually farmers help nature along by directing runoff. The ditches are called *larkakuna*, *zanjas*, *acequias*, or *mogote*. Ditches run up and down slope, but also deliver water to ditches perpendicular to the slope, along field boundaries, hedgerows, and semi-terrace risers, avoiding erosion of field surfaces. Stone diversion dams help shunt water into these perpendicular ditches.[36] Where a ditch delivers water to a terrace riser, a perpendicular notch in the riser (a "drop") forms. The drops may be plugged with hardpan blocks or have a stone at their base to limit erosion. In some cases, slope ditches will deliver storm runoff to an irrigation running along the contour of the slope, increasing the irrigation water supply of the canal users.[37]

It is clear that the elaborate slope drainage systems of Ecuador in part are the natural result of runoff along the edges of agricultural fields, coupled with ad hoc management to direct this runoff away from vulnerable crops. The resulting gullies may be deleterious in a number of ways, reducing soil moisture by lowering the slope water table, or constituting a hazard to people and animals.[38] In some instances, however, the slope drainage may improve soil workability, particularly where high clay content makes wet soils excessively slippery.[39]

Flat lands present somewhat different drainage challenges. In sandy flats such as San Roque (near Lake San Pablo) or around Lake Colta, drainage ditches are normally not necessary for farming, and are not

---

35 A rare example of stone faced, valley bottom, irrigated terraces can be found at the upper limit of La Rinconada (Angochagua) near Ibarra; potatoes, maize and other crops are grown. The features appear to be modern, and adjacent similar valleys lack them.

36 Such diversion dams can be examined on Loma Atallaro, near San Pablo.

37 Good examples occur along the Acequia del Pueblo of Pimampiro, on its upper course near Mariano Acosta.

38 For example, near San Bernardo in Chimborazo, children have fallen into the ten meter deep gullies between slope fields.

39 This appears to be the case in the Mira River drainage to the south of Pimampiro, where soils of high clay content have been carefully drained in prehispanic times; see Chapter 5.

present. However, fluctuations in lake level may create a hazard; in 1972, a rise in Lake San Pablo destroyed maize fields at San Roque Bajo.[40]

In flats with finer textured soils, such as those at San Pablo, Cayambe, and Zuleta, drainage is necessary to lower the water table and to prevent seasonal flooding. In all three cases, large farm owners assured me that the contemporary ditch pattern (1 meter deep ditches 150–200 meters apart) is adequate to protect pasture and crops, if the ditches are frequently cleaned. More closely-spaced ditches may be constructed by small farmers who do not have the luxury of being able to systematically manage the entire plain.[41]

Modern drainage ditches in flats need to be cleaned from 2 to 4 times a year, since they get clogged with vegetation (See Table 3.1) and fill with silt, eroded from ditch sides and the surrounding slopes. The *tarea de zanja* (ditch cleaning) takes about 7.5 person-days per 100 meters of ditch (6.7 person-days per cubic meter of material removed).[42] If ditches are spaced 200 meters apart, 37.5 person-days per operation per hectare of drained land may be involved, or an annual labor of 75 to 150 person-days per hectare; this is a substantial amount.

Irrigation and Its Functions

In the equatorial Andes, perhaps twenty percent of the sown area was under irrigation in 1985. This includes some areas of subtropical crops such as sugar cane and avocado, but the bulk of the irrigated area consisted of temperate region crops such as maize and potatoes.[43]

Irrigation can be practiced with a limited infrastructure. Fields located close to canals, springs, or with a high water table can be watered by hand.

---

40 PRONAREG-ORSTOM, *Diagnóstico, Documentos A–D,* Document B, 58.

41 More closely spaced ditches, about 15 to 20 meters apart, are occasionally constructed, e.g. by small farms at Alao. Perhaps the closer spacing assures small farmers that drainage will be achieved regardless of the drainage status of neighboring small farms. Fields on the Cayambe plain with similar ditch spacing appear to have been recently constructed (as landowners assured me they were), although it did not prove possible to ascertain their original purpose; for an argument that these features are prehistoric, see Bruce Batchelor, "Los Camellones de Cayambe en la Sierra de Ecuador," *América Indígena,* 40 (1980), 674 ff.

42 At Hacienda La Vega, it took 5 men with shovels three days to clean 200 meters of ditch (80 centimeters deep, 160 centimeters wide at the top), producing about 100 cubic meters of wet spoil in the process. This is the equivalent of 7.5 person-days per 100 meters of ditch. This is consistent with a report elsewhere that the task takes 8–10 person-days per 100 meters of ditch; see CIDA *Tenencia* , 272.

43 See INEC-MAG, *Sistema 1985*; this is the first attempt to reliably determine cultivated areas through a sampling procedure. Because irrigation is often supplemental and/or illicit, estimates of irrigated areas continue to be at best rough guesses.

Pots can be used to splash irrigate kitchen gardens.[44] By custom, gardens by the side of canals can be irrigated by those who otherwise have no right to the water, as long as the canal bank itself is not breached; water may be splashed out to the field or scooped into a temporary wooden or metal sluice for delivery to a ditch. Such practices are referred to by the verb *botar*, to (informally) splash out.[45]

In the flats, it is possible to derive water directly from the water table. An example is the *cochas* or sunken fields[46] of the Lake Colta flats; here pits have been dug down to the water table to enable cultivation of totora. Two harvests a year are possible, and the totora is used to make mats (*esteras*) and as cattle feed. Naturally marshy areas may also be modified and embanked for totora production, as at Lake San Pablo and Yaguarcocha.

Drainage systems can double as supplemental irrigation systems, particularly during the rainy season. The lower ends of slope drainage systems may feed into irrigated plots of onions or other crops, where sporadic runoff from downpours is concentrated and allowed to recharge upper layers of soil; the saturated soils will continue to allow maximum plant growth even during sporadic short term droughts in the rainy season.[47] In flats, drainage canals may also provide sources of water for irrigation during dry periods.[48]

Most irrigation however comes from canals and irrigation networks which transcend the abilities of individual families to construct and manage. Water for canals comes from both springs and rivers. The volcanic ash substrate is permeable, so that drainage density is low while the rivers and springs which do exist have relatively stable and permanent flows. Indeed, some rivers may have their maximum discharges in the dry season, due to the contribution of summer snow melt and spillover of precipitation from Amazon basin air masses near the headwaters.

The upper portions of the rivers flow through glaciated valleys with frequent flat-floored stretches where canals may be diverted via simple

---

44 See Rubio Orbe, *Punyaro*, 127, for splash irrigation near Otavalo.

45 These practices were observed in fields near Atuntaqui.

46 See Gregory Knapp, *The Sunken Fields of Chilca: Horticulture, Microenvironment and History in the Peruvian Coastal Desert*, Madison: Unpublished M.S. thesis, Department of Geography, University of Wisconsin, 1979; and Gregory Knapp, "Prehistoric Flood Management on the Peruvian Coast: Reinterpreting the 'Sunken Fields' of Chilca," *American Antiquity*, 47 (1982), 144–154.

47 For example, footslopes to the west of Colta Lake were irrigated in 1981 using the *canterón* method (see text, later in this chapter), relying on water only from slope drainage systems. This is an area which is generally considered to lack irrigation.

48 Because most flats are in the possession of large farms, this irrigation is now accomplished through employing gasoline powered pumps. A small sharecropped farm was observed, however, using water from drainage ditches in flats near San Pablo.

diversion dams (*tomas, bocatomas*). Lower down, the rivers flow through deep canyons eroded into volcanic ash fill. Flats on the volcanic ash suitable for irrigation frequently are 10 to 20 kilometers from the appropriate points for irrigation canal diversion. Thus community cooperation is required for irrigation of much of the equatorial Andes.

Canals are brought down to irrigated areas by a variety of technical solutions. Side ravines were crossed by means of aqueducts (*canales*) constructed of willow trunks supported by stakes; such aqueducts are said to have lasted more than 80 years before replacement.[49] In other cases, canals were brought under side valleys in tunnels topped with stone slabs (*socabones*). Canals were and still are brought through steep and difficult terrain by tunnels in the volcanic ash, cleaned via windows (*ventanas*).

If an irrigation system is in place and available to a farm family, the water generally is distributed via an intricate system of furrows. The configuration of furrows is continually managed by the use of the azadón as an all-purpose instrument of terrain modification. Perhaps the most characteristic furrow pattern is the *canterón* or serpentine furrow configuration, where the water zigzags down slope by plants which are growing from the furrow sides. This pre-Hispanic technology permits irrigation of very steep slopes without terracing.[50]

Labor requirements for irrigation are not as great as has been supposed, even when canal cleaning is included. Typically, canals are cleaned once or twice a year, with more frequent cleanings in dry zones. If the farmer does not help in this work, he must pay a day laborer to work in his stead. Approximately 160 person-days per year are needed to clean each kilometer of canal; in a typical irrigation zone, this may translate into 8 person-days per irrigated hectare.[51] Another requirement is the necessity for vigilance during water application; over an irrigation cycle of eight days, four hours a week may have to be reserved for the storage and distribution of water to a hectare. This is equivalent to 15 person-days (or nights) per hectare over a growing season of 8 months.[52]

---

49 Such a wooden aqueduct was used on the Acequia Alta of Ambuquí, according to local informants.

50 This discussion is based on a farmer's description of the process near Lake Colta. For a discussion of prehispanic serpentine furrows, see James Kus, *Selected Aspects of Irrigated Agriculture in the Chimú Heartland*, Ph.D. Dissertation,Ann Arbor: University Microfilms International, 1972, 178.

51 It is assumed that 8 person-days are needed to clean 100 meters of ditch in the characteristic volcanic ash soils (see section on drainage, above). Presidents of water boards were frequently questioned as to labor requirements for canal cleaning, and responses were generally consistent with this figure. If a typical zone has a 20 kilometer long canal supplying 400 hectares, each hectare would require 8 person-days of labor.

52 Commonly four irrigation hours are required per hectare of land per irrigation; see John K. Hatch, *The Corn Farmers of Motupe: A Study of Traditional Farming Practices in Northern Coastal Peru*, Madison: University of Wisconsin, Land Tenure

In practice, nocturnal water distribution turns (*turnos*) are not prized, and irrigation flow may go underutilized during some evenings. One adaptation to this problem is the practice of "entablar" a small hole is punched into the branch canal so that the fields fill slowly with water during the night.[53] Partly to enable more convenient timing of irrigation water applications, farmers on slopes may construct anks (*tanques*), excavated into the volcanic ash subsoil. In Pimampiro, tanks are about 5 by 10 by 3.5 meters deep. Water drains out via a tunnel in the subsoil, which is stopped up by a pole wrapped in cloth. The tanks have an added function of accumulating silt; up to 25 cubic meters of fertile silt may build up during the year, which can be excavated and applied to fields as a fertilizer.[54]

The silt produced from canals is called variously "lodo podrido," or "pozo negro." The most characteristic use of this fertilizer is after canal cleaning, but much of the material must also be delivered to fields suspended in irrigation water. The fertilizing qualities of silt accumulated in such tanks can be gauged by analyses of soil samples of silt from tanks, ditches, and canals (Table 3.4). Even where tank or canal cleanings are not available, the canals themselves bear suspended silt which, delivered repeatedly to fields, can replace nutrients lost to cultivation. Canals may be compared with streams, which collect substantial dissolved and suspended loads from their drainage basins. The Pimampiro canal, for example, runs below about 4 square miles of land above its course; 40,000 tons of sediment a year could be delivered to the canal, or 22,700 cubic meters,[55] enough to cover the 600 hectares it irrigates 0.38 cm thick with fresh material each year.

Some indication of this benefit can be seen in Table 2.1, in the nutrient comparison between irrigated and unirrigated fields.

The fertilization of irrigated fields with canal silt, coupled with the elimination of drought risk, results in increased yields. It has been

---

Center, Monographs Number 1, 1976. An eight day cycle is typical of northern highland Ecuador (e.g. Pimampiro). See Patricia Mothes, "La Acequia del Pimampiro: Riego Tradicional en el Norte del Ecuador," *Ecuador Debate*, 14 (1987), 69–86.

53 A practice described to me in the Ambuquí valley.

54 See Mothes, *Pimampiro's Canal*, 137–138 and 154.

55 The drainage area above the Pimampiro canal (below its offtake) is about 4 square miles, as measured by the author on topographic maps. Most of this area is cultivated or fallow land; little is in brush or forest. Soil textures are silty (loess). Sediment production of 10,000 tons per square mile may be a conservative expectation, based on soils of similar textures and uses in North America; see Thomas Dunne and Luna B. Leopold, *Water in Environmental Planning*, San Francisco: Freeman, 1978, 680, 682–683.

TABLE 3.4 FERTILITY CHARACTERISTICS, DITCH AND TANK FILLS

Site	Source/ laboratory	pH	Nitrogen (kg./ha.)	Organic Matter (%)	Phosphate (kg./ha.)
Pimampiro Tank A	Knapp/INIAP	7.4	116	—	120
Pimampiro Tank B	Knapp/INIAP	7.3	152	—	124
Pimampiro Tank C	Knapp/INIAP	8.1	88	—	193
Pimampiro Tank D	Knapp/INIAP	7.0	42	—	97
Pimampiro Tank E	Mothes/TAES	6.7	46	3.2	131
Pimampiro Tank F	Mothes/TAES	7.3	46	2.7	161
Pimampiro Tank G	Mothes/TAES	7.4	2	4.0	147
Pimampiro Tank H	Mothes/TAES	7.2	46	0.7	41
San Pablo ramp tola ditch fill	Knapp/SPAL	6.5	—	4.6	172
San Juan, Cayambe ditch fill (upper)	Knapp/SPAL	6.5	—	3.1	133
San Juan, Cayambe ditch fill (lower)	Knapp/SPAL	6.7	—	1.6	172
Milán, Cayambe tank fill	Knapp/SPAL	6.2	—	5.7	103
La Vega, San Pablo ditch fill	Knapp/ESPOCH	—	—	—	113
Chillogallo upper layer (0–11 cm.) ditch fill	Knapp/SPAL	6.2	—	6.9	78
Chillogallo upper ash fall (IIIC) (14–24 cm.)	Knapp/SPAL	6.4	—	2.2	28
La Tola ditch fill, 10–50 cm., Cayambe	Knapp/SPAL	5.4	—	3.8	23
Sigsicunga ditch fill, 0–20 cm.	Knapp/SPAL	4.9	—	10.1	30
Pimampiro canal fill Site I	Mothes/TAES	6.9	2.2	0.3	5
Pimampiro canal fill Site II	Knapp/INIAP	7.6	4.8	—	6

Key: INIAP: Instituto Nacional de Investigaciones Agropecuarias, Santa Catalina, Ecuador. SPAL: Wisconsin State Soil and Plant Analysis Laboratory, Madison USA. ESPOCH: Escuela Politécnica de Chimborazo Soil Laboratory, Riobamba, Ecuador. TAES: Texas Agricultural Experiment Station, College Station, Texas, USA.

estimated that irrigation commonly results in an increase of 150 percent of yields over those obtained in fields without irrigation or fertilizer.[56]

The Cultivation Process

The tool kit of agricultural tactics is deployed during the actual cultivation process. Regardless of crop, the steps include land preparation, sowing, weeding and care for the growing crops, harvest, storage, and food preparation. The details of these steps are determined by the crop.

Potato Cultivation

Since potatoes mature in the soil itself, potato cultivation places special emphasis on soil preparation and ridging; the crop also places special demands on harvest labor. The discussion of the major steps of cultivation which follows is based primarily on field interviews and observations in Chimborazo province.

1. INITIAL LAND WORKING (*TOLAR*). Regardless of whether the field is reclaimed from brush, grass, or simply field weeds, it is always necessary to turn over the earth (tolar) to allow the organic material on the surface to rot and bring up fresh soil from below. Often this step includes rastrar, breaking up clods.
2. MAKING FURROWS (*WACHUTA RURANA*) AND SOWING (*TARPUNA*). One person makes the furrow with the azadón while another places the seed potato and the fistful of dung. Furrows are made about 0.9 meter apart; this distance may be called the "wavelength" of the furrow set. Planters use their foot length as a gauge of seed spacing within the furrows; but the farmer's stated one foot distance corresponds in actuality to a 55 centimeter spacing, resulting in the characteristic Andean potato plant density of 20,000 per hectare. An average of 900 kilograms of seed potatoes are sown per hectare, although the value may be less in some areas.[57]

---

56 See Jack Keller et al., *Ecuador Irrigation Sector Review. USAID/Ecuador Water Management Synthesis Project Report 12*, Logan: Agricultural and Irrigation Engineering, Utah State University, 1982, 59. Measurements of irrigated and unirrigated fields by Mothes and Knapp (Table 2.1) indicate that unirrigated fields may have phosphate availability of 28 kilograms per hectare as compared with 98 kilograms per hectare in irrigated fields. This factor alone would create an increase of 1.55 in the fertility index.

57 The sierra average is about 900 kg./ha.; see PRONAREG-ORSTOM, *Diagnóstico, Documentos 1–11*, Document 4A. Communities in western Chimborazo reported

Terminology of the ridges and furrows is varied. Normally the term *wachu* (plural, *wachukuna*; Spanish huacho) is used for the furrows in which the potatoes are planted; the intervening ridges are called *lomas*. The lomas at this time are where people walk, so that an indented pathway (*calle*) develops on one side of the lomas. Later, as earth is heaped up over the developing potatoes, the original furrow becomes a ridge which is called wachu; the intervening furrows, previously ridges, remain the routes of transport and continue to be called calles.

There is substantial agreement on the labor input for furrowing and sowing, which may be estimated as 11 person-days per hectare with hand tools.[58]

3. WEEDING AND RIDGING. As the potatoes sprout and mature, earth is piled up around them to give the wachu a ridge form; this is claimed to enable the potato roots to develop better. Each ridging operation (called *aporque*, *kimina* Q.) increases the height of the wachukuna. Eventually, the distance from furrow bottom to ridge top (the "amplitude" of the ridge set) reaches 15–25 centimeters or more.

Ridges may be constructed such that they run up and down slope (appearing "vertical" when viewed from a distance); this practice is said to improve drainage on heavy soils. More commonly the ridges run close to the contour, appearing nearly horizontal when viewed from afar. Ridging is generally felt to reduce erosion and also to limit frost damage to developing plants.

The number of workings between sowing and harvest varies from three to five. Obviously, if the crop is doing poorly less effort is expended to than if a large harvest is expected. Each cultivation takes from 5 to 20 person days of hand labor, with the most onerous task being the final aporque (*colmar*). Total cultivation labor may be estimated at 60 person-days per hectare.[59]

4. HARVEST (*CAVAR*, *ALLANA* Q.). Potato harvest is hand labor, performed by several people flipping potatoes out of the soil with an azadón. Neighbors often contribute reciprocal labor at this time in exchange for part of the harvest. Few crops are as difficult to harvest as potatoes; they are buried in the ground, and their high moisture content and productivity may mean up to 20 tons must be

---

525 to 750 kilograms per hectare as plantings in marginal potato areas; see Cañadas, Salcedo, Bravo, and Knapp, "Vulnerability," 425.

58 Values elicited from farmers were close to survey results from small (0–1 hectare and 1–2 hectare) farms reported by PRONAREG-ORSTOM, *Diagnóstico, Documentos 1–11*, Document 4B.

59 Based on my judgment of the most reliable information from farmers in Chimborazo. Survey results for sierra cultivation of potatoes indicate a mean value of 53 person-days per hectare; see PRONAREG-ORSTOM, *Diagnóstico, Documentos 1–11*, Document 4B.

removed from each hectare. Harvest labor is proportional to the size of the crop, and portage is a substantial proportion of this labor. A harvest estimate of 10 person-days per hectare plus 3 person-days per metric ton harvested is consistent with much of the reported values as well as with field determinations for this study.[60]

5. STORAGE AND COOKING. Potatoes in Ecuador, unlike Peru, are not fermented or freeze-dried for preservation, but rather are kept in cool places in sacks. Since plantings are staggered in time and maturation times depend on elevation and variety, farmers with access to different fields either directly or through trade or barter can have fresh potatoes throughout much of the year. Potatoes are frequently sold, since prices are favorably high (there is no foreign competition with the local producer).

The major culinary use of potatoes is in soups and stews. Boiled peeled potatoes are also served as a side dish or as a complete meal; one typical highland farmers' lunch consists of a metal cupful of boiled potatoes. Labor inputs of the highland cooking process have yet to be determined.

Maize Cultivation

Maize cultivation is broadly similar to potatoes, but here the concern is to protect the exposed maize ears from predation and to exploit the relative ease of harvesting.

1. LAND PREPARATION. Since many maize fields are cropped continuously, labor inputs for preparation are usually less than for potatoes. The options for turning under organic matter are similar.
2. MAKING FURROWS AND PLANTING (*TARPUNA* Q). As with potatoes, maize furrows (*wachukuna*) may be made either with plow or with azadón. Furrow wavelength is similar to that for potatoes: about 0.75 to 1.0 meters. Holes are punched into the furrow with a digging stick about 40 to 90 centimeters apart; 1 to 4 maize seeds are thrown into each. Often a couple of beans are added, and squash (*zambo*, *zapallo*) is also sown. A variety of seeding densities are possible, from 10 to 40 kilograms per hectare; the average for the equatorial Andes is about 29 kilograms/hectare.[61]

---

60 For 1–2 hectare farms in the sierra, mean potato yields were 4.3 metric tons, while harvest labor was reported at 22 person-days per hectare; see PRONAREG-ORSTOM, *Diagnóstico, Documentos 1–11*, Documents 4A and 4B. In earlier reports I suggested use of the value 6 person-days per metric ton harvested.

61 Calculated by the author from PRONAREG-ORSTOM, *Diagnóstico, Documentos 1–11*, Document 4A.

3. LAND CULTIVATION (*JALLMANA* Q.). In contrast to potatoes, land in maize is typically worked only once or twice. In the process, the land is either worked up into ridges (wachukuna) similar to potato ridges, or into individual hills. Labor input may vary from nothing to 30 person-days per hectare.
4. HARVEST. Even maize which has been poorly cultivated or afflicted with drought or disease can be used as cattle fodder. Usually the maize ears are removed from the standing maize, and later the maize stalks are cut down and gathered for fodder, or, occasionally, left in the fields for grazing, incorporation into the soil, or burning. Labor costs for removing the ears are low, perhaps 10 person days per hectare. Maize may be harvested early as corn on the cob (*choclo*); or it may be harvested later as mature corn.
5. STORAGE AND PREPARATION. Maize harvested as choclo is boiled and eaten fresh, sold to marketers, or marketed directly by the farmer at periodic markets or roadside stalls. The rest of the maize is stored in the rafters of peasant houses. Maize kernels must be removed from the cobs; the corn may then be toasted or ground for a great variety of traditional dishes. Probably it is most common to incorporate maize into soups or stews, either as choclo or as flour; maize is also toasted, baked (*humitas*), popped, and converted into a drink (*chicha*). All these methods are relatively labor intensive, although labor inputs are not now known.

Barley Cultivation

Unlike most of the native Andean crops, barley (*cebada, sibada* Q.) and wheat are not planted individually but rather sown broadcast. There is no elaborate ridging and furrowing of the planting surface, and harvest is performed with a sickle. Barley therefore requires fewer steps in cultivation and lower labor inputs.

1. LAND PREPARATION AND SOWING. Barley is seldom planted as the first crop in a newly prepared field; it rather is characteristic of lands with medium to short fallows, as part of a rotation sequence with potatoes, maize. and/or legume crops (e.g. broad beans) to help restore soil nitrogen. A smooth planting surface is prepared, with labor inputs close to those for maize, and approximately 80 kilograms/hectare of seed is sown broadcast.[62]
2. FIELD CULTIVATION AND HARVEST. After sowing, the grain fields are subject to problems of sheetwash and gullying, as the

62 Seed application rates were remarkably consistent for the communities interviewed; see Cañadas, Salcedo, Bravo, and Knapp, "Vulnerability," 424.

unridged surface does not block or absorb flowing water. Care is minimal; labor inputs for small grain cultivation are far less than for any other major highland crop Total labor inputs for barley cultivation can be estimated as 54 person-days per hectare.

Labor inputs for major highland crops are summarized in Table 3.2.

Rotations, Fallows, and Yields

If land reclamation costs are higher for cultivation on brush or grass-covered land, yields are also higher due to the higher soil fertility. Field interviews are currently the only source of information on fallowing practices and relationships with yield.

1. POTATOES. On very recently reclaimed land, potato yields are high. A large landowner with lands on high slopes near Cayambe reported yields of 36 metric tons per hectare on land reclaimed by *pudrición*, with a second year crop of 18 metric tons per hectare. In the Colta area, informants reported 9 to 13.5 tons/hectare on lands reclaimed from grass, with drops to 7 metric tons the second year and 5.5 tons the third year. Since yields decrease with continuous cropping, many farmers resort to fallowing to restore soil fertility. Short fallows (*samachina* Q.) may be from 2 to 6 years in length; this is long enough to restore fertility but not long enough for large woody shrubs to develop; cattle and sheep graze the invading herbs. One source[63] claimed that typical initial yields after samachina in good years are about 4.5 to 9.0 tons per hectare; a second crop would be about 2.7 tons per hectare, and a third crop 1.8 tons, but this is usually considered not worth the trouble.
2. MAIZE. As with potatoes, maize may be cropped on reclaimed land. An example is the steep slopes above Hacienda La Vega near San Pablo, where a cultivator claimed that a reclaimed field yielded 3 metric tons per hectare the first year; despite heavy manuring, the yield dropped to half this value in a few years. In this same region, older maize fields on steeper slopes are rested from 1 to 4 years between harvests; land in rest is called *tampu* and the practice of short fallowing is called here *purma*. Land on gentler slopes and near settlements is almost continuously cropped for maize. Without fertilization, such maize yields about 210 to 450 kilograms per hectare, depending on slope and soil quality;

63 This information is from an indigenous farmer with extensive experience in a region (western Chimborazo) where it is very difficult to get reliable yield information.

with ample animal manure, frequently cropped fields yield 900 to 1,500 kilograms per hectare.[64]

3. BARLEY. Barley is seldom planted on fresh land; rather, potatoes or maize are planted and barley is introduced as yields decline. Barley is in fact the normal crop planted after potatoes in a short fallow situation; the barley crop is followed by broad beans, fallow, or another potato crop. Typical yields of highland barley fields are about 700 kilograms per hectare, with an estimate of 428 kilograms per hectare for unfertilized barley following a potato crop.[65]
4. ROTATION AND INTERCROPPING. In both potato and maize areas, rotation is commonly practiced. The most common sequence is to begin with potatoes, which are heavily fertilized. After the potato harvest the field is considered *puebla*, not fertile enough for potatoes but with enough residual fertility for a less labor-intensive crop such as barley. If barley is to be followed by another crop, a nitrogen-fixing legume such as beans, peas, lentils, or (especially at higher elevations) broad beans is chosen. This sequence of potato-barley-legume, so characteristic of higher elevations, results in declining yields in calorie and protein terms[66] but is a workable alternative to leaving the field in complete fallow for animal grazing.

---

64 These estimates of maize yields are based on field estimates late in the 1980–1981 growing season (a relatively good year). A reliable local guide helped in the estimation of probable yields in fields near San Pablo based on visual observation of height of corn and size of cobs.

65 Average highland small-farm (1–2 hectare) barley yields in the mid 1970s were reported as 712 kilograms/hectare; see PRONAREG-ORSTOM, *Diagnóstico, Documentos 1–11*, Document 4A and 4B. The Galte-Chauzán communities reported yields of 826 kilograms per hectare with manure and 428 kilograms per hectare without manure in a wet year; the same survey showed that barley was never cropped in two successive years, but most commonly was followed by broad beans; see Cañadas, Salcedo, Bravo, and Knapp, "Vulnerability," 424. Philip Westra, agronomist for the Luke Society, very kindly provided me with the results of his field measurements of barley production on the west side of Chimborazo province in the good year of 1981; yields of 70 fields varied from 90 to 3185 kilograms per hectare, with a mean of 1,092; mean measured soil fertility of his plots was relatively high at about 132 kilograms per hectare nitrogen and 75 kilograms per hectare phosphate. Since these fields were from farmers interested in receiving technical help, they probably are on average better than the highland norm.

66 The characteristic yields of broad beans are about 430 kilograms per hectare without fertilization and 680 kilograms per hectare with fertilization, in a context where broad beans are following barley in a rotation cycle; this is about 28 percent of the calorie yield and 79 percent of the protein yield of the average preceding barley crop. Yield data is from Cañadas, Salcedo, Bravo, and Knapp, "Vulnerability;" calorie and protein conversions from Woot-Tsuen Wu Leung, *Tabla de Composición de Alimentos para Uso en América Latina*, Mexico City: Interamericana, 1964.

5. SECTORAL FALLOWING AND CULTURAL LANDSCAPES. Unlike parts of the southern Andes, community cooperation in fallow and rotation management (sectoral fallowing) does not appear to be practiced in Ecuador. Distinctive cultural landscapes are however created in various regions by the co-evolution of cropping practices and livestock management between neighboring farms. Thus in a region of potato cultivation, the logical sequence of potato/small grain/broad bean/fallow will create a field patchwork of these various crops interspersed with fields with fallow grazing; the harmony which appears planned is in fact the product of individual adaptations to similar selective pressures. Nowhere is this principle more evident than in the question of hedgerows or field boundaries; once one tradition of animal management is established in the landscape, most fields tend to correspond to its norms. In some regions, animals are tied down to graze on field margins; the result is a landscape of fenceless fields with grassy margins.[67] In other cases, cattle graze freely in fenced fields; the landscape is a maze of hedgerows and adobe, *tapía*, or stone walls. Where irrigation is practiced, field margins are designed to facilitate irrigation as well as animal control; crops correspond with what the local marketing infrastructure makes most profitable.

In another context, Enrique Mayer has written of Andean "production zones" as delimited "bounded ecosystems" with "precise boundaries" which can be readily mapped, and have "a characteristic type of field" which exhibit a "specific form of social organization."[68] Such a characterization could be applied to Ecuador only with the cautions that (1) since group controls outside irrigation are relatively weak, the visible production zones are epiphenomena of adaptive processes operating largely at the level of the individual farms; (2) group decision-making is itself highly varied in range and institutionalization, so social organization cannot readily be extrapolated from one production zone to another which superficially resembles it; and (3) production zones are internally heterogeneous, subject to stress and adaptation, and subject to both historical change of the zone as a whole and to historical redefinition of production zone boundaries themselves. In

---

67 Especially good examples of these open, unfenced landscapes for maize production can be found on the eastern slopes of the volcano Imbabura.

68 Mayer suggests there is a characteristically Andean tradition of the supra-individual organization of society, nature, and space. For an explanation of the idea of production zone and the use of this idea to define a project of characterization of Andean cultural geography, see Enrique Mayer, "Production Zones," in Shozo Masuda, Izumi Shimada and Craig Morris (eds.), *Andean Ecology and Civilization*, Tokyo: University of Tokyo Press, 1985, 45–84. Mayer argues that where group determinants of agricultural practice are less obvious (e.g. at lower elevations), this is due to the decay of previous group controls.

other words, production zones have no ontological or explanatory status as superorganic or holistic entities or systems, apart from the processes of contextual adaptation and choice which result in the (often transitory and even illusory[69]) partitioning of the environment into apparently different landscapes.

Intensive, Continuous Cropping

In a few situations it is possible to find farm plots which are multiple cropped, with two harvests a year and high productivity. Such a farm located in the drained flats of the San Pablo wetland was studied to indicate the limits of highland productivity.

This small share-cropper plot had been cropped for the six preceding years by a wife and husband. The owner and the farmers were interviewed, and the farm was visited several times. The farm is located between two ditches. Although the farmer does not contribute to the maintenance of these ditches, they are nevertheless important for drainage and irrigation. The pastures contain abundant prehistoric raised fields (see Chapter 7).

The farm is about 50 meters square (0.25 hectare), mostly pasture. About 0.03 hectare is house garden and cultivated field. A one meter deep ditch and corresponding bank next to the farm maintain the water table below 50 centimeters of the field surface. The bank nurtures a typical range of useful hedgerow plants (Table 3.1, plus *Physalis peruviana* and *Solanum caripensis*). The farm itself is surrounded by a bank with barbed wire fence strung between stakes of the echero tree. A canal crosses the farm on the way to a shallow cattle pond; this is useful for irrigation during dry periods. Canal water also supports useful stands of marsh reeds and edible plant, as well as nitrogen-fixing blue-green algae.

The farm house is surrounded by a small kitchen garden and yard with lechero trees, tree tomatoes (*Cyphomandra betacea*), and several clumps of culinary herbs.[70]

The house and its garden are flanked by two fields. One is used for potato-maize double-cropping and the other for potato double cropping.

---

69 If, as is argued here, labor and land efficiencies are powerful determinants of Andean agricultural practice, environmental boundaries which affect efficiencies would be expected to encourage agricultural adaptations accordingly, but with exceptions due to the specific conditions (and ignorances) of each farmer.

70 Capulin cherry (*Prunus capuli*), guaba (*Inga* sp.), *babaco* and *chihualcán* (*Carica* spp.) and *tacso* (*Tacsonia* spp.) are also common fruits grown elsewhere in these flats.

The two fields combined contain 0.018 hectare. According to the owner, they are fertilized with every other year with 600 pounds of cattle manure gathered (free of charge) from nearby pastures. This constitutes 16,667 kilograms of manure per hectare per year, or 8,333 kilograms (not dry weight) per crop. The presence of canal and drainage ditch also means that the field is fertilized with ditch cleaning spoil and suspended nutrients in irrigation water.

The basic tool for all work is the *azadón*. Both fields are worked and then sown in August. Potatoes are sown in furrows with handfuls of manure. At the same time, lines of broad beans (*habas*) and clumps of peas, quinoa and alfalfa are sown.[71]

After a few weeks, the land is weeded, and then the earth is heaped around the potatoes to form the distinctive wachukuna or 1 meter wide potato ridges. The wachukuna reach an ultimate height of 15 centimeters.

In October or November, maize, beans, and squash are planted in the furrows between the ridges in the maize-potato double cropped field. A few maize seeds are also planted in the other field. Potatoes are harvested from both fields in January, and replanted in the potato-potato double cropping field.[72] The maize is ridged into wachukuna to prevent wind throw. In the year of observation, this maize was harvested for green corn in April, although it could have been left to ripen. The second crop of potatoes is harvested in June or July.

This farm is said to produce, in an average year, 225 kilograms of potatoes,[73] 45 kilograms of green corn, 11 kilograms of beans, and 3 kilograms of quinoa, in addition to squash, broad beans, and peas. Less potato seed, the system produces about 15 million calories/year/hectare; this would support about 1,500 persons per square kilometer.[74]

On the other hand, labor requirements are clearly large; they may be estimated at 226 person-days per hectare only for the farming tasks.[75] The

---

71 Tarwi is common in the area but not sown on this particular farm.

72 In the year of observation, the farmer chose to plant peas instead of potatoes as the second crop in this field.

73 The yield is equivalent to 9 tons/hectare potatoes.

74 Assuming one million calories per person. Calorie equivalents are taken from Leung, *Tabla*.

75 The quantity of potatoes produced is equivalent to about 10,000 kilograms/hectare, so 10 + (3 x 10) or 40 person-days/hectare would be required for harvest. For the 40 percent of the area in potato double-cropping, labor inputs are 20 (weed reclamation) + 11 (sowing) + 60 (cultivation) + 40 (harvest) for each crop. The maize crop requires 20 + 10 + 30 + 10.

farm is dependent on the presence of ditches, which require periodic cleaning, at about 7.5 person-days per 100 meters cleaned. This labor is dependent on ditch spacing; if ditches are 20 meters apart in one direction and 100 meters in the other, to facilitate irrigation, then 90 person-days of labor per year are needed.[76] Also, labor is needed to manage and watch irrigation, possibly involving about 40 person-days a year per hectare. If a family of five is supported on 0.33 hectare, about 117 person-days per family would be needed using this system.

76 This particular farm family, as sharecroppers, were not required to participate in ditch cleaning activities.

Chapter 4

MODELS, SCENARIOS, AND THE LARGER CONTEXT

Modelling Adaptive Strategies

Enough information has now been presented to model agricultural yields and labor inputs under varying assumptions of fallow length, rotation, and irrigation. The model will be set up in such a fashion that when a value for overall population density (P) is entered, the model will calculate the unique fallow length, crop yield, and labor input per family which will sustain that density. The use of population density as the independent variable is justified by earlier work on farmer adaptive dynamics.[1]

In this chapter, the model will be presented with inputs and parameters characteristic of Ecuadorian highland small farms in the mid 1970s. The same model will be modified in Chapter 8 to reflect assumed characteristics of pre-Hispanic farms. In both cases, the model results will be checked against other information.

The Inputs and Parameters

For small farms in the sierra in the 1970s, the model will be based on the following values.

1. The yield constant (y) in tons per hectare. This is the yield of an optimally fertilized plot during a year of median precipitation at the

[1] For example, Ester Boserup, *The Conditions of Agricultural Growth*, Chicago: Aldine, 1965. See also B. L.Turner II and William Doolittle, "The Concept and Measure of Agricultural Intensity," *Professional Geographer*, 30 (1978), 297–301; H. C. Brookfield, "Intensification and Disintensification in Pacific Agriculture: A Theoretical Approach," *Pacific Viewpoint*, 13 (1972), 30–48; and Peggy F. Barlett, "Labor Efficiency and the Mechanisms of Agricultural Evolution," *Journal of Anthropological Research*, 32 (1976), 124–140.

site in question. In this chapter, the values used will be based on optimal yields at the Plaza Farms for each precipitation value.

2. The yield correction factor (x), a dimensionless constant. This adjusts for the need to store food for years with yields less productive than those of median precipitation. A value below 1 reduces the available yield by that amount, which goes into storage to provide for suboptimal years. In areas similar to the Plaza farms, where the diminished yields of years with above-median precipitation is balanced by the increased yields of years with below-median precipitation, an assumption of no storage in a typical year is probably workable, and x can be set at 1.
3. The calories per ton of crop (c). This will be set at 3.61 million for maize, 0.79 million for potatoes, and 3.48 million for barley.
4. The seed requirement per hectare (s). This will be set at 900 kilograms for potatoes, 36 kilograms for maize, and 79 kilograms for barley.
5. The number of harvest calories needed to support a person per year (d), and number of persons per family (p). Since highlanders consume about 1,733 calories per day,[2] a minimum of 0.63 million calories per year per person are required for subsistence. Today, however, highland small farmers purchase some of their food with earnings from cash sales and off-farm work. The 1974–1975 socio-economic survey of highland farms suggested that small (1–2 hectare) sierra farms or UFAs (family agricultural units) produced about 1.75 million calories each, and this value will be used here for the family calorie need.[3]
6. The amount of nitrogen fertilizer available per family (n) will be estimated as the production of typical family farm (65 kilograms) multiplied by 40 per cent (the amount usable by plants), or 26 kilograms. Of this total we will assume about half (or the night-time production) is actually used, or 13 kilograms.
7. The amount of phosphate fertilizer available per family (f) will be estimated as the production of a typical family farm (28 kilograms) multiplied by 60 per cent (the amount plants can use), or 17 kilograms. Again we will assume about half is used.

---

2 The average calorie consumption of all individuals above 1 year of age in three villages in northern Pichincha province, north of Quito, was 1,769 calories per day; for Peguche, near Otavalo, the value was 1,697 calories per day. Averaging these two values yields 1,733 calories per person. See Vargas and Gallegos, *Diagnóstico*.

3 See PRONAREG-ORSTOM, *Diagnóstico, Documentos 1–11*. There were in 1974–1975 about 77,716 highland farms in the 1–2 hectare category; 55,998 hectares were actually cropped, 41 percent in maize, 24 percent in small grains, 11 percent in tubers, 9 percent in legumes, and 15 percent in other crops. A rough estimate of total calories produced per farm, based on standard tables, is 1.75 million calories.

8. Labor requirements for cultivation (L_c) will be estimated as 30 person-days per hectare for maize and 60 person-days per hectare for potatoes. Harvest labor (L_h) will be estimated as 10 person-days per hectare for maize and 10 person-days per hectare plus 3 person-days per ton for potatoes. Irrigation labor (L_i) will be estimated as 21 days for potatoes, 28 days for maize.[4]

The Basic Model

The model is based on seven simultaneous equations, based on the constants and parameters already defined and solving for the following seven unknowns: the length of fallow in years (r), the initial soil phosphate in kilograms per hectare (F'), the initial soil nitrogen per hectare (N'),[5] the amount of usable added nitrogen in kilograms per hectare (N"), the amount of usable added phosphate in kilograms per hectare (F"), the yield per hectare in tons (Y), and the land area cropped per family in hectares (h). The equations are listed in the footnote to this paragraph.[6]

The solution of the unknowns allows for further calculations of labor input and manure application. It is assumed that reclamation labor (L_r) is a function of fallow length (r) according to an equation developed earlier in this study.[7] With this calculated value, it is then possible to determine the total labor used for agriculture per family(L_t):

$$L_t = (L_i + L_c + L_h + L_r) / h$$

---

4 On the basis of 8 days for canal maintenance, plus 4 hours a week during the growing cycle for irrigation management and vigilance, both per hectare.

5 Measured initial nitrogen was not incorporated in the fertility index equation developed at the Bejarano farms. However, in order to indicate the combined effects of phosphorus and nitrogen depletion on yields, some estimate of nitrogen as a function of fallow needs to be made. Here, the logarithmic regression equation of nitrogen against fallow time developed in Chapter II will be used: $N" = 84r^{0.13}$. Since the Bejarano fertility index equation was developed assuming an initial nitrogen value of zero, and assuming a fallow time of 4 for the Bejarano farms, the initial nitrogen equation for use in fertility index calculations becomes $N' = 84r^{0.13}-100.8$, where 100.8 is the value of the equation at r = 4.

6 The equations are $r = (5/Ph)-1$; $F' = 21r^{0.4}$; -100.8; $N' = 84r^{0.13}-100.8$; $N" = n/h$; $F" = f/h$; $Y = yx\,(1-.506(10-.011(N")))(1-.726(10-.0027(F'+F")))$; and $h = pd/(y-s)c$. These equations are either logical consequences of the definition of the unknown or are based on data developed in this study and are discussed in the text. The equation relating initial soil phosphate to fallow length was developed in Chapter 2.

7 The equation is $L_r = 20.1r^{0.28}$.

Results for Manured Potatoes

Table 4.1 indicates the results of runs of this model for the case of potatoes, under incremental assumptions of overall population density, and for a median precipitation year at the Plaza farms.[8] As expected, as population density increases, fallow times and yields decline. Labor inputs per family also decline, in keeping with the assumptions of agricultural intensification theory. At a population density of 92 families per square kilometer, there is no solution to the equations; it is impossible to feed this number of people under the assumptions of the model. The minimum feasible fallow length is about 1.7 years, and the minimum feasible yield about 6.3 tons. At this level of productivity, manure applications are about 6.7 metric tons per hectare.

TABLE 4.1 MODELLED POTATO YIELDS AND LABOR

Overall population density (UFA* per km.^{2})	Fallow, years	Yield (tons/ha.)	Labor (person-days/ family)	Manure (tons/ha.)
2	364	17.1	31	61
32	13	11.0	32	23
40	9.7	10.3	33	21
48	7.4	9.8	33	18
56	5.7	9.3	35	16
64	4.5	8.8	36	14
72	3.6	8.2	37	12
80	2.8	7.6	39	10
88	2.0	6.8	43	8
90	1.7	6.3	46	6.7
92	*Not possible to support this population density*			

*UFA = Family agricultural unit

[8] In this and all other cases, model results were calculated by an iteration procedure using a computer spreadsheet program.

The Rotation Results

Table 4.2 shows the results of running the model for the same median precipitation under the assumption that unfertilized barley is planted the year after potatoes. The formulae for initial phosphate, nitrogen, and reclamation labor inputs are adjusted to reflect the shortening of the fallow cycle by one year, and the labor inputs and yields are adjusted upwards correspondingly.[9]

It can be seen that with modelled rotation, a greater population density can be supported (112 families/hectare). Modelled labor requirements per family are lower for unrotated fields at low population densities, but when population densities pass 80 (and fallow lengths decline to below 3 years), it is more labor efficient to rotate.

TABLE 4.2 MODELLED POTATO YIELDS: ROTATION

Overall population density (UFA per km.^{2})	Fallow, years	Yield (tons/ha.)	Labor (person days/ family)	Manure (tons/ha.)
2	404	17.3	34	75
16	39	13.4	34	46
32	17	11.9	35	36
48	9.8	10.9	36	31
64	6.6	10.1	38	27
80	4.6	9.3	39	23
96	3.3	8.5	41	19
112	2.0	6.9	47	13
114	*Not possible to support this population density*			

The Maize and Irrigation Results

Table 4.3 shows the results of the model for maize, for a median precipitation year at the plaza farms. Again, with increased population density, modelled fallow lengths and yields decline and modelled labor inputs per family increase. The modelled strategy is able to support a maximum of 104 families per square kilometer; at this point, modelled fallows are about 1.4 years long, yields are about 1,200 kilograms/hectare in median precipitation years, and about 7.0 tons of manure per hectare are

9 It is assumed that each year farmers cultivate an equal area of potatoes and barley, and that barley requires 50 person-days a hectare to cultivate (including reclamation) and yields 428 kilograms/hectare but with a seed requirement of 79 kg/ha.

applied. Maize growing is more labor efficient than potato growing. Maize fallows can also be reduced somewhat further than potato fallows.

TABLE 4.3 MODELLED MAIZE YIELDS AND LABOR

Overall population density (UFA per km.^{2})	Fallow, years	Yield (tons/ha.)	Labor (person days/ family)	Manure (tons/ha.)
8	72	2.9	18	39
64	4.7	1.8	19	15
72	3.8	1.7	20	14
80	3.1	1.6	21	12
88	2.5	1.5	21	11
96	2.0	1.4	22	9
104	1.4	1.2	25	7.0
106	*Not possible to support this population density*			

Table 4.4 indicates the results of modelled maize irrigation, based on the assumptions of a yield constant of 6.48 tons, and a correction factor of 1.0.[10] It is assumed that labor inputs per hectare are 8 person-days for irrigation canal maintenance and 15 person-days for field irrigation. It is also assumed that irrigation improves soil fertility in a cumulative fashion, so that initial soil phosphate is increased by (20/(r+1)) x 2.4.[11]

Under irrigation, the result is an inverted labor input curve: labor efficiencies actually increase at higher intensification levels. Labor inputs per family for irrigated maize are extremely low, even with the conservative assumptions about increased yield due to added water.

---

10 The assumption of a 63 percent increase of median precipitation yields over those of the Plaza farms (at 94 millimeters per month, 1,032 millimeters per year) is based on the regression equation of precipitation against yield established by Table 2.6, and an assumption of irrigation cultivation at a site with 700 millimeters/year mean annual precipitation.

11 This equation is based on an assumption of 0.38 centimeters/year deposition of irrigation-borne silt, with a silt fertility similar to the average of irrigation tanks at Pimampiro, e.g. 126.7 kilograms/hectare. Each year, then, the 20 centimeter thick topsoil layer is improved by 0.38/20 x 126.7 = 2.4 kilograms/hectare of phosphate. The factor 20/(r+0.1), where r is length of fallow, indicates the number of silt deposits in the preceding 20 years, which are assumed to cumulatively contribute to fertility.

TABLE 4.4 MODELLED IRRIGATED MAIZE YIELDS AND LABOR*

Overall population density (UFA per km.^{2})	Fallow, years	Yield (tons/ha.)	Labor (person days/ family)	Manure (tons/ha.)
20	50	5.0	12	119
120	6.1	4.2	11	83
240	2.4	4.0	11	75
360	1.2	3.9	10	73
480	0.7	3.9	10	73
600	0.33	3.9	10	72
640	*Not possible to support this population density* **			

*Assuming 700 millimeters a year mean annual precipitation

**Fallow under 0.25 years

Manipulating the Models

An advantage of modelling is that it permits the exploration of alternative assumptions.

1. FERTILIZATION. The model is very sensitive to changes in family manure supply. If no fertilizer is available, the maximum potato carrying capacity falls to less than half its former value, and fallows are longer and yields are less. Labor inputs also increase for a given population. Increasing fertilizer supply has the inverse effect, increasing carrying capacities, reducing possible fallow times, and lowering labor inputs.
2. CHANGES IN YIELD CONSTANT AND CORRECTION FACTOR. The reduction in the median precipitation yield and the reduction in the amount of the harvest that can be used for current consumption (correction factor) have the same effect on the model. For example, if the corn yield constant is reduced by 25 per cent (to 3 tons), the carrying capacity is reduced by 46 per cent (to 56 families/square kilometer), and labor inputs increase while yields decline. Table 4.5 summarizes the modelled population density at carrying capacity for varying assumptions of maize yield constant.

Modelling Population by Crop and Precipitation

As mean monthly growing season precipitation increases above 94 millimeters, modelled yields decline. Table 4.5 shows the effect of declining precipitation on yield constants and modelled population carrying capacity.[12] It is not easy to relate the mean monthly precipitation during the growing season to mean annual precipitation. A median mean monthly growing season precipitation of 94 mm for maize may correspond with about 1,030 millimeters per year,[13] in which case the information in Table 4.5 shows that by about 1,500 millimeters a year maximum maize zone population densities have dropped to less than one family per square kilometer.

TABLE 4.5 ABSOLUTE MAXIMUM POPULATION DENSITIES BY PRECIPITATION, MAIZE

Median monthly precipitation (mm.)	Yield constant (tons/ha.)	Carrying capacity, (UFA/km^2)
94	4	104
106	3	56
118	2	26
130	1	9
136	0.5	3
139	0.25	0

The maximum values in Table 4.5 are achieved with progressively higher labor inputs as yield constants go down. Table 4.6 indicates population densities at different populations levels if labor inputs per family are held constant (the hypothesis of labor equality). If labor inputs are held constant at about 25 person-days per family, only the area around 94 millimeters monthly precipitation is used to capacity; by 106 millimeters population density drops to zero, as no level of intensification is labor competitive with the favored zone. This is consistent with a limitation of maize farming to an area with median annual precipitation of 1,160 millimeters or less. Potatoes are not grown, as they cannot compete with maize in terms of labor efficiency.

At values of precipitation lower than 94 millimeters a month, yield constants continue to increase, but are counterbalanced by increased risk of

12 The relation between precipitation and yield is that indicated by Table 2.6.

13 Median maize growing season precipitation was about 90 millimeters per month. Most maize sites are near the stations Hacienda Angla, which has a median annual precipitation of 988 millimeters (mean 915 mm). A median of 94 millimeters might then correspond with (94/90)(988) = 1032 millimeters per year. Median growing season precipitation for potatoes was 94 millimeters; median annual precipitation for average interpolated Angla-Zuleta values (the values most often used for potato sites) was 1025 millimeters. (mean 1065).

TABLE 4.6 POPULATION DENSITIES BY PRECIPITATION AND CROP, ASSUMING COMPARABLE LABOR EFFICIENCIES

		Calculated UFA densities per km.^{2}			
		Marginal labor input = 25 person-days/UFA.*		*Marginal labor input= 50 person-days/UFA**	
MEDIAN PRECIPITATION (MM./MONTH)	YIELD CONSTANTS, MAIZE/ POTATOES	MAIZE	POTATOES	MAIZE	POTATOES, UNROTATED
94	4/19.6	104	0	104	112
106	3/17.2	0	0	56	58
118	2/14.7	0	0	15	26
130	1/12.3	0	0	0	5
142	0/9.8	0	0	0	0

*At these labor inputs, irrigated maize would support a density of 600 at 700 millimeters mean annual precipitation.

loss to drought.[14] Modelled overall population densities remain constant near this optimal precipitation value and to slightly below it, before beginning to decline due to rising soil moisture deficits in median years.[15]

Implications: A Modern Subsistence Scenario

The modelling procedure has produced a series of data of land and labor efficiencies for alternative agricultural tactics. The data are consistent with limited field interview information and have the advantage of consistency and transparency of derivation. The data can in turn be used for the analysis of the implications of various scenarios of farmer decision-making. Here, the focus will be on a scenario where small-farm labor inputs are roughly equalized.

---

14 Yield adjustment factors fall as median precipitation nears the optimum for the crop concerned; under these circumstances, yield under median precipitation is much greater than median yields.

15 Plausibly, the observed lower precipitation limit of maize cultivation of about 600 millimeters is in response to processes similar to those described here. Correlations of maize yield with precipitation developed in Africa suggest an asymmetrical curve, with more rapid yield declines at the dry end than at the wet end of cultivation; the optimal yield value is closer to the dry limit than to the wet limit of cultivation. See Glover, "Relationship."

It has been assumed that all production goes to subsistence. If it is further assumed that consumers have no preference for potato, barley, or maize calories, then an effort-minimizing farmer should employ only that strategy or mix of strategies which uses labor at least as efficiently as the alternatives for the prevailing population density. The key indicator is labor input per family; and farmers should attempt to use the lowest-labor alternative available to them under prevailing population densities. If they are socially excluded from certain zones, they should strive to optimize labor efficiencies via the opportunities open to them.

It should be emphasized that the purpose of the scenario is to explore the implications of certain assumptions. In the real world, obviously no single factor ever dominates adaptive choice, and no scenario can ever be expected to be fully realized. Nevertheless, the predictions of scenarios can be checked against the real world, and the results can help identify actually operative processes, constraints, and selective pressures. Such is the aim of the present study.

If population densities are very low, modelled irrigated maize farming requires less labor than any competitor for producing the family subsistence; only 10 person-days per family suffices to provide food for a year (Table 4.4). Due to the inverted labor curve, even at low population densities the irrigated niche is used to capacity.

At higher population densities, irrigated maize remains labor efficient; indeed, only when all the potentially irrigated land is used up do other strategies appear competitive by this measure. Unirrigated maize farming is next to irrigated farming in labor efficiency, especially in more favorable climatic locations. In a scenario of effort minimization, as population densities become fairly high in the most favored maize locations, less favored locations should become attractive at longer fallow intervals, together with potato farming.

This modern subsistence scenario thus predicts several patterns at higher population densities.

1. YIELDS. At high population densities, all the available strategies should be used, albeit at differing levels of intensification. If population densities have reached 112 families per square kilometer in the areas with potato/barley production (50 person-days per family), the irrigated maize areas should be fully used (about 600 families per square kilometer), the better-watered maize areas pushed to capacity (104 families per square kilometer), and the less-advantaged maize areas intensified. Yields of maize at optimal precipitation levels should be about 1,200 kilograms per hectare, declining in areas of greater rainfall; similar yields of potatoes should be about 6,900 tons per hectare (barley rotation). Even at high levels of overall labor input, potatoes are only feasible if they produce many more calories per hectare than maize.

2. ALTITUDINAL ZONATION OF CROPS. Since it is usually labor inefficient to grow potatoes where maize will grow, potatoes should be restricted to (1) zones too high and cold for maize, or (2) zones where potatoes' shorter maturation time allows more optimal crop timing with wet or dry seasons, e.g. where precipitation is very high (Table 4.6). Note that this explanation does not depend on any presumed change in yield or pest sensitivity of potatoes with respect to elevation, but rather on the ecological limits of maize and the differential labor input characteristics and yield:seed ratios for the respective crops.[16]
3. CHARACTERISTIC FERTILIZATION RATES AND FALLOW CYCLES FOR MAIZE AND POTATOES. This scenario suggests that rotated potatoes will be fallowed longer and fertilized more than maize. It also suggests that at higher population densities potatoes will be rotated with barley. Again, these implications are not based on any inherent differences in the ecological requirements of the respective crops.
4. PRECIPITATION CORRELATION WITH A SINGLE CROP ZONE. The scenario suggests that the maize zone should not have uniform crop practices, but that fallow times, manure rates, population densities, and crop yields should vary continuously from place to place, with the highest yields and population densities in areas with the optimum climate: the least chance of crop loss due either to excessive or deficient precipitation. The same should be true of zones of potato cultivation. Rotation practices should also vary from place to place in accord with precipitation, or indeed any environmental factor affecting yields. If intensification is allowed to proceed to the point where 50 person-days per family is allowable, then the precipitation "bandwidth" of agriculture widens for maize (to perhaps the equivalent of 1,290 millimeters a year), while the precipitation bandwidth of potatoes extends to about 136 millimeters/month (1,420 mm/year).
5. IMPACTION. Potentially irrigable zones should be used fully, with little or no fallow, and planted to maize below its feasible upper elevation limit. Since per family labor inputs are still quite low when the irrigable zone is used to capacity, there should be efforts to further intensify agriculture within the zone through elaboration of cultivation techniques and practices, and searches for possibilities to expand irrigation. Such maximal intensification has been called impaction by Geertz, although with the

[16] The ability of potatoes to support slightly higher populations than maize at the limit of intensification might suggest that at very high population levels potatoes would "invade" maize zones. Except at higher precipitation levels, however, the differences between the two crops are too close to permit any confidence in this speculation.

presumption that labor inputs were high rather than low.[17] Indeed, the entire area where intermediate precipitation leads to good maize yields should "fill" to capacity fairly readily.

6. CHALLENGES OF DIFFERENTIAL LABOR EFFICIENCIES. The tendency for areas suitable for maize agriculture to fill to capacity should lead, as population densities increase, to attempts to gain some access to land in other areas (e.g. potato lands) to employ additional family labor, albeit less efficiently. At the same time, the coexistence of production areas with markedly differing labor efficiencies provides a problem and a challenge for society: how should these niches be allocated in such a way as to minimize or control conflict between haves and have nots?

The scenario may be further adjusted by altering the assumptions as to farmer motivation. If farmers sell crops on the market and purchase food, while remaining self-sufficient as to fertilizer, land, and labor, this has the effect of changing the calorie values of the model; now the calories associated with each crop are those the crop is able to purchase if sold on the market. In today's Andes, where grain imports depress maize and barley prices, the advantage is often ceded to potatoes, which "yield" much more than maize. Thus potatoes can be predicted to be grown in zones suitable for maize if the object is sale, and if potato prices are stable and high.

Finally, if labor is only intermittently available, for sowing and harvest but not for cultivation, the cultivation labor factor can be greatly adjusted upward to reflect its scarcity. Under such conditions, barley can come to seem even more attractive as an option, perhaps even without rotation with potatoes. This corresponds with the situation where family members seasonally migrate to the urban economy.

Checking the Scenario

Many of the implications of the scenario discussed above seem to correspond to observable patterns in the contemporary Andes. Such correspondences are very encouraging indicators that some underlying processes active on the Andean scene have been identified.

In 1982, the population density of the potentially arable equatorial Andeswas about 140 persons per square kilometer. This is the population of the equatorial Andean provinces (2.7 million, excluding Santo Domingo de los Colorados) divided by the total area between 2,000 and 4,000 meters elevation and with between 500 and 2,000 millimeters a year precipitation

---

17 Geertz, *Agricultural Involution.*

(19,100 square kilometers). Population density outside the provincial capitals was about 70 per square kilometer.[18]

For the Ecuadorian highland provinces as a whole, 77,716 0–1 hectare farms (UFAs) in the mid 1970s controlled about 949.1 square kilometers, for a total of 82 family farms per square kilometer.[19]

This is quite a high population density, near the maximum modelled carrying capacity of many strategies except for irrigation. Checks of other elements of the scenario provide reassuring indications that the modelling procedure is realistic.

1. YIELDS. Average maize yields determined by mid-1970s interviews in small highland farms averaged 617 kilograms per hectare; reported potato yields averaged 4,305 kilograms per hectare. Given the fact that these include areas of varying and often suboptimal precipitation, these values are low but quite in line with model predictions. The spread between potato and maize yield (and especially the calorie productivity of potatoes versus maize) is even greater than predicted, perhaps due to the tendency to shift manure resources to potatoes in view of the latter's greater marketability.
2. ALTITUDINAL ZONATION. Potatoes are progressively displaced by maize in lower elevations suitable for maize production, as has been demonstrated by quantitative analysis of land use as a function of elevation in the vicinity of Ibarra.[20] Although this pattern could also be explained by presumed disease factors, the latter have never been empirically demonstrated and it is at least suspicious that the presumed disease vulnerability begins precisely where shortened maize maturation times permit maize agriculture.
3. FALLOWS. The actually cropped area of 1–2 hectare farms in Ecuador was approximately 59 per cent of the total area.[21] The implied fallow is longer than expected for irrigated maize, about the same as expected for rotated potatoes, and somewhat greater than expected for unirrigated maize. In general, the model predictions seem reasonable.
4. PRECIPITATION CORRELATIONS. The suggestions that the upper limits of maize and potatoes should be near 1,300–1,400 millimeters isohyet seems consistent with observations of crop

---

[18] As measured on the map by PRONAREG-ORSTOM (Programa Nacional de Regionalización Agraria – Office de la Recherche Scientifique et Technique Outre-Mer), *Ecuador: Mapa Ecológico. 1:1,000,000*, Quito: Instituto Geográfico Militar, 1978.

[19] PRONAREG-ORSTOM, *Diagnóstico, Documentos 1–11*, Document 2, 102.

[20] Gondard, *Inventario y Cartografía*, 63.

[21] PRONAREG-ORSTOM, *Diagnóstico, Documentos 1–11*.

limits in Chimborazo province, where few crops are planted in areas with over 1,400 millimeters mean annual precipitation.[22]

Calendars

Golte has claimed that in the Andes "the sum of the labor requirements for the agricultural cycles in the different altitudinal zones represents a much more balanced curve, and, of course, a much greater quantity of labor days in production itself than in other agricultural societies." He suggests that "the explanation" for the high levels of Andean productivity supporting great empires is this calendrical efficiency, what he calls "Andean organization" of land and labor.[23]

Golte develops a quantitative measure of calendrical inefficiency: seasonal unemployment, defined as the ratio between the total labor time spent on a crop and the amount of time which would have been spent if all months were as busy as the busiest month. Use of this measure implies that during the busiest month farmers are working at the level they prefer, and that during the rest of the year farmers are frustrated, unable to work to their liking because of the lack of appropriate tasks. Obviously, agricultural intensification can only proceed to the point that the busiest month fully employs the laboring population.

Using Peruvian data, he suggests that seasonal employment is higher if only potatoes or maize are grown than if both are grown together. This suggests that use of multiple crops (which Golte equates with use of multiple altitudinal zones) can increase carrying capacity.

A number of conceptual and methodological problems with Golte's approach have been pointed out elsewhere.[24] Here it is worth indicating that very high population densities can be supported with relatively low family labor, and that in no case is the theoretical maximum family labor input per month (30 x 5 or 150 person-days) even remotely approached; some of the techniques reach their carrying capacity at *annual* labor inputs less than this. Since plantings are staggered in distinct *suertes*, both planting and harvest labor are spread over several months—labor inputs are "elastic," in Golte's terms. Furthermore, farmers generally prefer discontinuous over continuous labor inputs, since discontinuous inputs provide large blocks of free time which can be used for social and non-agricultural economic activities.[25] Thus crops with very high indices of

---

22 Cañadas, Yugcha and Estrada, "Effects on Zones."

23 Golte, *Racionalidad*, 25–27. The translations are mine.

24 Knapp, *Soil*, 202 ff.; Gregory Knapp, *Ecología Cultural Prehispánica del Ecuador*, Quito: Banco Central del Ecuador, 1988, 107–109.

25 See Boserup, *Conditions*, 53–55; Golte, *Racionalidad*, 49–50, recognizes the importance of this point.

seasonal unemployment (barley) have in fact been preferred by farm families exploiting other off-farm opportunities.

The Larger Economy and Society

This discussion so far has emphasized models and scenarios of small-farm behavior based on the assumption of decision-making at the family level of organization: least-effort employment of resources to meet basic family goals of subsistence and reproduction. Scenarios under these apparently unrealistic assumptions produce results which are in some cases surprisingly close to reality, suggesting explanatory power. The scenario, however, suggests two areas where individual families need to call on the larger community to solve key problems. Firstly, the superiority of irrigation over other strategies suggests the need for canal construction and maintenance, tasks which inherently require group cooperation. Secondly, the fundamental inequalities of labor efficiencies between niches requires some arrangement for niche allocation which reduces or eliminates the potential for conflict between differentially privileged households.

Ecological Complementarity, the Hacienda, and the State

Some niches, such as irrigated maize areas, are relatively labor efficient even at carrying capacity. The feeding of larger populations than these niches could support implies either that (1) some families are disadvantaged, requiring more labor expenditure than other families, or (2) families equalize labor expenditure by sharing both attractive and unattractive niches.

The second solution constitutes verticality or ecological complementarity and can be considered part of an "explanation" of these patterns. Interestingly, the use of very dry or very wet niches within the maize zone would be an alternative to the use of potato zone niches to achieve equality.

This second solution has nothing to do with putative desires for dietary or crop diversity or calendrical efficiency but rather is a simple response to population growth in a diverse environment under a constraint of equality. This constraint of equality can be produced by the attempts of a group to reduce conflict by minimizing intergroup differences.

Elites can also exploit frictions in peasant populations due to differential labor efficiencies across niches. An elite could appropriate the favored niches and farm them under a different mode of production than in the less favored niches. This corresponds in part with the hacienda system, although obviously is not a sufficient explanation of it.[26] Verticality and the hacienda can thus be considered to be different ends of the same continuum, competing and opposite ways to control conflict over Andean niches.

These brief points may serve as indications for pathways of ecological analysis of Andean institutional history. It is not, however, claimed that Andean social institutions can be predicted or reduced to epiphenomena in terms of adaptive and ecological factors alone, because multiple equally suitable solutions exist for given problems

Irrigation Organizations

The extraordinary superiority of maize irrigation over other options provides a powerful incentive for the formation of social units to build and maintain irrigation canals. It is becoming increasingly clear that there is no pan-Andean mode of irrigation organization today, but rather a diversity of organizational tools and strategies. In addition to the immediately visible organizational forms, aspects of group identity and ethnicity have been called upon from time to time to help expand, defend, or take over irrigated areas.

At present the agency responsible for the control of irrigation water and the construction of new irrigation works is INERHI. In practice, the actual maintenance of canals and control of irrigation water is in the hands of local water boards (*juntas de agua*), agricultural associations, cooperatives, communities (*comunas*), groups of family gardeners, and agencies of the government.[27]

Prior to 1972, and in accordance with the 1939 water law, water rights could be transferred independently of land (the "Yemenite" principle of water law).[28] After the 1972 water law, water could no longer be transferred independently of land, but inequalities in the distribution of water are recognized as valid.

Apart from the rivers on the eastern flank of the Andes, the rivers of the equatorial Andes are considered fully utilized for irrigation.[29] Despite this, there are continued efforts to find water sources, plan new projects, and obtain funding. Wherever political power is concentrated, irrigation

---

26 Alternatively, small farmers could be permitted to control and farm all the niches, but subject to differential rent, tribute, or taxes based on the differential labor efficiencies.

27 Keller et al., *Ecuador Irrigation.*

28 See Thomas Glick, *Irrigation and Society in Medieval Valencia*, Cambridge: Harvard University Press, 1970; and Mothes, *Pimampiro's Canal.*

29 See for example PRONAREG-ORSTOM (Programa Nacional de Regionalización Agraria – Office de la Recherche Scientifique et Technique Outre-Mer), *El Agua con Fines de Riego. Provincias de Carchi e Imbabura*, Quito: Ministerio de Agricultura – ORSTOM, 1982.

projects will be on the agenda, whether it be a hacienda, indigenous organization, water board, parish, or development project.[30] The great expense of such projects does not discourage attempts to receive outside funding.

Two recent examples give an idea how canals have been undertaken with varying types of organization.

1. THE CANAL OF LA VICTORIA was constructed near Atuntaqui in the middle of the nineteenth century, when farmers of Atuntaqui contracted with José Manuel Jijón owner of the Hacienda Pinsaquí near the offtake location. The canal was destroyed by the 1868 earthquake; a new canal was constructed which, after numerous litigations, passed to the control of Atuntaqui in 1912 and continued in use up to the present by a population accustomed to the political exploitation of its mestizo identity.[31]
2. THE MAGDALENA CANAL, near Caranqui, had its origin with the donation in 1903 of several springs on Hacienda Magdalena; the canal was constructed by farmers of the Caranqui area and benefitted both small farms and hacienda land. Almost immediately there was litigation over the propriety of taking water which previously flowed into a river which supplied other canals. The legal result was a condemnation of the spring donation but no specific action was undertaken to stop canal use, which continues to the present.[32]

Indigenous Organizations

In general, outside irrigation associations, small farmers in Ecuador are weakly organized. Alliances are sought through real and fictive kinship, but there are few opportunities for village-wide or regional action. Whether the result of the last few centuries of hacienda domination, or a heritage of earlier patterns, today peasant communities play a minimal role in conditioning household decision-making, and most of the more valuable land is held as private property.[33]

---

30 Farmers of the Evangelical Indigenous Association of Chimborazo were interested in using both their organization and outside development personnel to bring irrigation water to the area of Colta Lake. In the Pimampiro area, citizens, hacienda owners, and farmers are often very well informed about the various speculative possibilities of using regional rivers.

31 Pedro M. Zumárraga, *Monografía del Cantón Antonio Ante*, Quito: La Prensa Católica, 1949 220–223.

32 "Revindicación de las aguas donadas por la familia Gómez de la Torre...," (1905), Archivo Histórico del Banco Central, Ibarra, Juzgado 1º, No. 94, 202 ff.

33 See PRONAREG-ORSTOM, *Diagnóstico, Documentos A–D*, Document A, 87; CIDA, *Tenencia*, 166; Costales and Costales, *Katekil*, 76; Rubio Orbe, *Punyaro*,

State Policy Against Agriculture

By far the most important factor conditioning contemporary small farmer decision-making has been the policy of the Ecuadorian state to directly or indirectly suppress prices by importing foreign grain. Wheat imports to Ecuador increased from 118,000 metric tons in 1973 to an estimated 380,000 metric tons in 1985. The wheat has been imported on favorable credit terms, and prior to 1975 was accompanied by the provision of free grain as food aid. In tandem, production of wheat, barley, and soft corn all peaked in late 1960s in Ecuador and have decreased since.[34]

The record of provision of help to small farmers has been spotty. There is no national extension service in Ecuador to help small farmers, and until very recent times there was no reliable information gathering system to determine crop production. Development projects by the government, private organizations, and internationally-funded agencies have provided useful background studies and specific help to certain locations, but have suffered from uncertainties of funding, changes in direction, and lack of follow-up on completed projects.

28–29, 106. See also Weismantel, *Food*, 53: the "lack of social organization beyond egocentered kin and compadrazgo networks contrasts with the classic image of the Andean community." Weismantel observes that the *minga* is seen as a tool of hacienda or state rather than the community, and that the *comuna* simply functions as a subdivision of the parish rather than an alternative.

34 In recent years barley production has recovered somewhat. See USDA, *Ecuador Situation Report 1987.*

Chapter 5

PREHISTORIC AGRICULTURAL ADAPTATIONS

The crops and technology in use today are in part the heritage of pre-Hispanic cultures, but it would be a mistake to see Andean agriculture as a simple continuation of indigenous practices. In fact, five of the eight major subsistence crops of the Andes were brought by the Spaniards,[1] along with most of the domestic animals and many of the tools, as well as the overall economic and political framework.

There is evidence that the Spanish introductions were voluntarily accepted by indigenous farmers with explosive rapidity,[2] so that while the contemporary complex is not pre-Hispanic, it does in essence date to the first decades of Spanish control. This rapid acceptance is in keeping with the advantages of the new crops as discussed in the preceding chapter, as well as with the fundamental contribution of the European animals to the manure supply.

At the same time, certain technical alternatives were abandoned after the Spanish conquest. These include raised fields, certain kinds of terracing, and the crops although not the technology of subtropical irrigation. To understand these lost technologies, evidence must be sought in early accounts, archives, and in the landscape itself.

The Inca and Before

The influence of the brief period of Inca control on the equatorial Andes was relatively slight, especially from the perspective of agricultural technology, perhaps being limited to the introduction of squash and

---

1 Maize, potatoes, and beans are indigenous; barley, wheat, habas, peas, and lentils introduced.

2 "...Knowing the benefit and utility of wheat and barley, many of the natives subject to this city of Quito sow them, consume them and make grogs of barley," Pedro de Cieza de León, *La Crónica del Perú [1553]*, Madrid: Espasa-Calpe, 1962, 130–131, translation mine.

camelids.[3] Most of the surviving agricultural landforms as well as much of the native testimony in the archives refer to the situation under various chiefdoms which dominated the equatorial Andes prior to Spanish arrival.

Chiefs and Chiefdoms

Prior to the Inca conquest, the equatorial Andean peoples were politically organized into chiefdoms, societies without a specialized administrative apparatus but with considerable personal power concentrated in the hands of powerful chiefs; indeed, the chiefdoms were often known by the names of their leaders.

Chiefs often had several wives, and families of up to at least 42 persons; in one case, 7.3 percent of a community population was included in the family complexes of various leaders. They also commanded the labor of "attached" families, servitors (*yanakuna*), and corvée labor. The goals of the chief's activity included subsistence of the people directly concerned, but even more so the provision of large ceremonial occasions of consumption, which cemented the chief's power.[4]

Chiefdoms were "nested," with smaller sub-chiefdoms with lesser chiefs owing some allegiance to larger units with more paramount chiefs. The smallest unit was the group or *parcialidad*, as the Spanish called it; the parcialidad was headed by an ethnic lord (*principal*). The parcialidades were organized into larger indigenous communities which the Spanish called *pueblos de indios*; the heads of the pueblos were referred to as chiefs (*curaca, cacique*). The pueblos were in turn sometimes under the control of a dominant pueblo headed by a paramount chief.[5] Ties between different groups could be facilitated by marriage, or by the custom of sending the

---

3 See Frank Salomon, *Los Señores Etnicos de Quito en la Epoca de los Incas*, Otavalo, Ecuador: Instituto Otavaleño de Antropología, 1980, 256, and Chantal Caillavet, "La Adaptación de la Dominación Incaica a las Sociedades Autóctonas de la Frontera Septentrional del Imperio (Territorio Otavalo-Ecuador)," *Revista Andina*, 3 (1985), 403–423. For the suggestion that squash was introduced by the Inca see Sauer, "Cultivated Plants," 504.

4 Anonymous, "La Cibdad de Sant Francisco de Quito," in Marcos Jiménez de la Espada (ed.), *Relaciones Geográficas de Indias - Perú, Volume 2*, Madrid: Biblioteca de Autores Españoles, Ediciones Atlas, 1965, 226; Salomon, *Native Lords*, 124–142; Salomon, *Señores Etnicos*, 198, 202.

5 For chiefdom political geography in highland Ecuador, see Chantal Caillavet, "La Estructura Básica de las Sociedades Autóctenas de la Sierra Norte del Ecuador en el S. XVI," Conference on Reproduction and Transformation of Andean Societies of the Social Science Research Council, Quito, 1986; Salomon, *Native Lords*, and Horacio Larrain Barros, *Demografía y Asentamientos Indígenas en la Sierra Norte del Ecuador en el Siglo XVI*, Otavalo: Instituto Otavaleño de Antropología (Two Volumes), 1980; and Horacio Larrain Barros, *Cronistas de Raigambre Indígena*, Otavalo: Instituto Otavaleño de Antropología (Two Volumes), 1980.

child of one chief to be raised in the family of another. It is possible that relations between different principales and chiefs were normally close to equality (symmetrical), with war being associated with greater emphasis on dominance and subordination.[6]

The basic parcialidad units which made up the pueblos were not necessarily contiguous with them. There is evidence of ethnic lords and subject groups being sent to fairly distant locations for the cultivation of specialty crops, at least by Inca times.[7] In general, however, the various parcialidades had definite territories and boundary zones, and were geographically contiguous with the dominant pueblo.

Salomon has suggested using the term *llajta* (plural *llajtakuna*) to refer to the pueblos, and the term "ethnic lord" to refer to the principales and caciques.[8] This procedure will be followed here unless it is desired to point out the hierarchical position of a particular llajta or lord; in the latter case, the terms parcialidad, pueblo, dominant pueblo, principal, chief, and dominant chief will be used.

Ramp Tola Sites

Various statements in the sixteenth century geographic reports allow us to picture the chiefdom landscape. The most imposing house in each llajta was the house of the ethnic lord. This was built by the people, and was the site of big feasts and ritual drinking. The most valued lords were those which had the best-cultivated fields in the vicinity of their house. The fields were worked by their subjects. There is evidence that yields per unit seed were higher on the ethnic lords' lands.[9]

---

6 Compare Salomon, *Señores Etnicos*, 205–206, 212; and Larrain Barros, *Demografia*, Volume 1, 118–119.

7 See, for example, the case of the parcialidad/pueblo of Axangui, located close to Perucho but politically subject to Otavalo, apparently located so as to exploit local opportunities of cotton irrigation. Sancho de Paz Ponce de León, "Relación y Descripción de Los Pueblos del Partido de Otavalo [1582]," in Marcos Jiménez de la Espada (ed.), *Relaciones Geográficas de Indias - Perú; Volume 2*, Madrid: Biblioteca de Autores Españoles, Ediciones Atlas, 1965, 233–242, and Caillavet, "Estructura Básica."

8 Frank Salomon, *Native Lords*.

9 See Antonio Bello Gayoso, "Relación que Enbio a Mandar Su Magestad Se Hiziese Desta Ciudad de Cuenca y Toda Su Provincia [no date]," in Marcos Jiménez de la Espada (ed.), *Relaciones Geográficas de Indias - Perú, Volume 2*, Madrid: Biblioteca de Autores Españoles 184, Ediciones Atlas, 1965, 271 and 286; Anonymous, "Cibdad," 226; and Salomon, *Señores Etnicos*, 195–196.

Where the llajtakuna were part of larger political units, the lord of the llajta had a house in both the local ethnic territory and in the vicinity of the chief.[10]

Ethnic lords in the areas north of Quito frequently constructed their houses on top of mounds, called *tolas*.[11] These tolas began to be constructed sometime between A.D. 200 and A.D. 700. A special type of rectangular mound with a ramp leading to its flat top (ramp tola) was constructed after about A.D. 1250; the tops of ramp tolas frequently are of the order of 200 to 600 square meters in area. Excavations of the large mounds and ramp tolas have indicated that most if not all were house platforms; troughs are often present in the tops of the mounds, possibly serving for the preparation of maize beer or as hearths for cooking ritual meals.[12]

Tolas and ramp tolas are typically grouped in sites with numerous mounds. Zuleta, the largest site, has 148 mounds, 13 with ramps. The tolas and tola sites of the northern highlands have been mapped with the aid of air photographs, and appear to correspond with the political centers of pre-Hispanic parcialidades,[13] with each tola corresponding to an ethnic lord's family. Unfortunately, there is no clear correspondence between size and number of tolas and ceramic characteristics with the ethnohistorically known population or importance of the llajtakuna, except that tola sites which correspond with pueblos seem usually to have ramp tolas.[14]

---

10 Larrain Barros, *Demografía.*, Volume 1, 179–180; Salomon, *Señores Etnicos*, 194–195.

11 "Tola" appears to be a term taken from the pre-Inca language of the zone of Otavalo. A chief's son's last will and testament referred in 1592 to a "tola" where his father used to have his house; see Caillavet, "Toponimia," 10–11.

12 See Athens, *Evolutionary Process*; Wolfgang Wurster, "Aportes a la Reconstrucción de Edificios con Planta Circular, Sobre las Pirámides con Rampa de Cochasquí," in Udo Oberem (ed.), *Cochasquí: Estudios Arqueológicos*, Otavalo: Instituto Otavaleño de Antropología, 1981, 79–124.

13 As also suggested by Caillavet, "Estructura Básica."

14 The tola sites of the northern highlands were first mapped by Fernando Plaza Schuller, *Contribución al Estudio de los Montículos Prehistóricos de los Andes Septentrionales del Ecuador: Apuntes de Aerofotointerpretación Arqueológica*, Otavalo: Instituto Otavaleño de Antropología (Manuscript Atlas), 1977. This atlas maps each individual tola. A later atlas which maps tola sites rather than individual tolas is Gondard and López, *Inventario Arqueológico*. The details of the Zuleta mound site are from Athens, 262.

Farms and Settlement Patterns in Chiefdoms

Although perhaps 20 percent of a llajta population could be associated permanently with the households of chiefs and principales,[15] the rest of the population operated "family farms" which produced maize and potatoes for their own subsistence. In general, these households stored their own maize and potatoes for use throughout the year; they also manufactured maize beer (chicha) which was drunk on various ceremonial and festival occasions, to cement family alliances, so that annual production exceeded strict subsistence requirements.[16]

The households were fundamentally made up of nuclear families, only rarely with lateral extensions. The juridical domestic census units in a 1559 visita south of Quito had an average of 4.06 persons each, although the presence of units which clearly could not have been households suggests that the de facto household size was somewhat larger, perhaps at least 5 persons.[17]

The houses of the common people were typically small, round huts with walls of lashed poles covered with mud (*bahareque*) and roofed with thatch.[18] A small house could be built in a few days.[19] Early sixteenth century Spanish observers noted that the bulk of the Indian population was dispersed, with little nucleated settlement.[20] Archaeology tends to indicate that this was also the pre-Hispanic pattern; there is slight evidence of commoner settlement between tolas, and indeed settlement sites located away from tolas are normally small.[21] It can be concluded that people normally lived close to their fields.

Population

Archaeological evidence, accounts of early observers, and the reports of numbers of tributary Indians all are consistent with the idea that highland population was dense in pre-Hispanic times and declined abruptly during

15 For example, in the llajta of Uyumbicho, the households and servitors of the chiefs and principales included 19 percent of the total population; see Salomon, *Native Lords*, 120.

16 Salomon, *Native Lords*, 76–77.

17 Salomon, *Native Lords*, 16, 119–120.

18 Paz Ponce de León, "Relación," 240.

19 Anonymous, "Cibdad," 226.

20 There were few "pueblos poblados en forma;" see Anonymous, "Cibdad," 224.

21 No pre-Hispanic commoner house foundations have yet been found in the northern highlands. Settlement sites are identified by ceramic scatter and presence of manos, metates, and bones. The largest settlement sites are the defensive terraces above the irrigated area at Pimampiro, but these are quite unusual in scale.

the sixteenth century. Later the reconstruction of population densities in pre-Hispanic times will be discussed in greater detail.

Crops and Diet

There is little doubt that the basis of the pre-Inca diet was maize. Maize was consumed as beer (*chicha, aswa*), toasted kernels, flour, popcorn, and probably also as corn on the cob (*choclo*) and hominy (*mote*).[22] Maize was consumed as part of household meals, as well as in ceremonial contexts.[23] Other seed crops included kidney beans, chile pepper, domesticated lupine (*tawri, tarwi, chocho*), and quinoa. Tubers included, most importantly, the potato, along with the other Andean tubers, uka, melluku, and mashua; jíquima and sweet potato (*camote*) were grown at lower elevations.[24]

Fruit trees were grown in highland valleys in prehistoric times, particularly in the lower elevations. The list would include avocados, *lúcuma* (*Lúcuma obovata*), *grenadilla* (*Passiflora* sp.), highland papayas (*chihualcán, babaco*), *chirimoya* (*Annona muricata*), tree tomato (*Cyphomandra betacea*), guaba, pepino, and guayaba. Also below about 2,500 meters cotton would be grown, and below 2,000 meters, manioc, peanuts, and coca.[25]

---

22 See Anonymous, "Cibdad," 226; and Lope de Atienza, "Compendio Historial del Estado de los Indios del Perú [1575?]," in Jacinto Jijón y Caamaño (ed.), *La Religión del Imperio de los Incas*, Quito: Escuela Tipográfica Salesiana, 1931, 52–53.

23 Maize was consumed at sowing, at harvest, at a summer festival, during the rite of first hair cutting, during the male puberty rite, and during the funeral rites, when maize beer was buried. In all these cases, the giving and receiving of maize involved the creation or reproduction of social ties, obligations to work, and personal loyalties. See Salomon, *Señores Etnicos*, 128–131.

24 See Cieza de León, *Crónica*, 121, 130, 131; Anonymous, "Cibdad," 226; Paz Ponce de León, "Relación," 235 and 237; and Diego Rodríguez Docampo, "Descripción y Relación del Estado Eclesiástico del Obispado de San Francisco de Quito que se Ha Hecho por Mandado del Rey Nuestro Señor en Virtud de su Real Cédula Dirigida al Illmo. Sr. D. Agustín Ugarte Saravia...[1650]," in Marcos Jiménez de la Espada (ed.), *Relaciones Geográficas de Indias - Perú, Volume 3*, Madrid: Biblioteca de Autores Españoles 184, Ediciones Atlas, 1965, 75; Bello Gayoso, "Relación," 286.

25 See Sauer, "Cultivated Plants," 521–532; Paz Ponce de León, "Relación," 238–240; Fr. Antonio de Borja, "Relación en Suma de la Doctrina e Beneficio de Pimampiro [1582]," in Marcos Jiménez de Espada (ed.), *Relaciones Geográficas de Indias - Perú, Volume 2*, Madrid: Biblioteca de Autores Españoles 184, Editorial Atlas, 1965, 249–250; P. Joel O.M. Monroy, "El Convento de la Merced de Quito de 1534 a 1617 (1a Parte)," *Boletín de la Academia Nacional de Historia (Quito)*, 10 (1930), 159; Fr. Gerónimo de O.M. Aguilar, "Relación fecha por mi, Fray Gerónimo de Aguilar ... de la Doctrina y Pueblo de Caguasquí y Quilca...[1582]," in Marcos Jiménez de la Espada (ed.), *Relaciones Geográficas de Indias - Perú, Volume 2*, Madrid: Biblioteca de Autores Españoles 184, Ediciones Atlas, 1965, 245–247;

Some of the subtropical and tropical crops, notably cotton, chili pepper, and coca, were obtained through centralized markets, through trade by specialized traders (*mindaláes*), and through trade by specialized householders through periodic trading trips. These trade networks covered quite large territories. The institution of placing ethnic colonies at sites of important subtropical resources (*kamayuj*) was probably an Inca innovation.[26]

Animals

As was the case with crops, indigenous people avidly accepted the introduced Spanish animals: chickens, pigs, sheep, and cattle. Many households developed animal management practices, including the use of corrals, early in colonial times, but the full use of the high grasslands for sheep production was a slower process, as indigenous farmers introduced a mixed subsistence economy and the larger landowners developed the woollen textile industry, based on flocks in the high páramos and grassy valley bottoms cared for by full-time shepherds.[27]

The aboriginal animal economy was much simpler and indeed more impoverished. In general, meat was only eaten during the principal feasts. The various animal meat sources, including guinea pig, did not contribute much to the general diet. Indeed, the only meat sources listed in 1650 by Rodríguez Docampo were "some fat grubs which grow in the ground, called cusos," and a "little fish...called choncho" (*Pimelodus cyclopum*?).[28]

The ethnic lords had special privileges in eating game; part of the meat was handed over to the chiefs, while specialized hunters worked for some lords.[29]

Camelids were present in northern Ecuador before the Inca but their use was limited. Even at the end of Inca rule, camelids near Otavalo were mentioned only as sumptuary meat sources for ethnic lords; near Quito, they are mentioned only as pack animals. Further south, in areas under Inca domination for a longer time, camelids were used for wool and for sacrifices, but the technology was probably introduced by the Inca, for state use, and vanished as rapidly as it appeared.[30]

Jorge Garcés (ed.), *Libro de Proveimientos de Tierras, Cuadras, Solares, Aguas, etc., por los Cabildos de la Ciudad de Quito 1583–1594*, Quito: Archivo Municipal Publication XVIII, 1941, 101.

26 An extensive discussion of transfers of tropical and subtropical goods to the highlands can be found in Salomon, *Native Lords*, 97–115.

27 Galo Ramón Valarezo, "Del Cacicazgo Andino a la Hacienda: La Transformación del Espacio Productivo en Cayambe," *Cultura (Quito)*, 8 (24b) (1986), 639–654.

28 Rodríguez Docampo, "Descripción," 75.

29 Salomon, *Native Lords*, 83.

30 Salomon, *Native Lords*, 85–87, 190.

Fertilization

The general absence of domesticated animals in pre-Hispanic times suggests that household manure sources were limited to people, dogs,[31] and guinea pigs. In the back rooms of houses they had "great crowds of … guinea pigs," fed with "quantities of grass; and the amount of their droppings [*estiércol*], together with the grass which covers it, is such that the smell is unbearable."[32]

Colonial documents indicate a close relationship between manuring and cultivation, particularly of the higher elevation slopes and poorer soils. Thus, in an archival document of 1647 cited by Ramón,[33] the seven principal ethnic lords of Cayambe litigated for possession of a herd of sheep, arguing that for many years they had rights to manure: "taking advantage of the aforesaid sheep they manure their fields of potatoes, barley, and other crops … and … without the aforesaid sheep they could not cultivate since the land is bitterly cold and borders on the páramo." Another document of 1686 refers to lands which "only could be sown by those who had much manure with which to fertilize them." There are also references at this time to sheep corrals walled with sod blocks and sown with broad beans, maize, and potatoes.[34]

In prehistoric times, manure resources would have been restricted. The household manure resource was that available from humans and guinea pigs; perhaps about 1140 kilograms per family per year, equivalent to 33 kilograms nitrogen and 16.1 kilograms phosphate (Table 3.2).

Cultivation

Virtually the only early (1573) extended description of agricultural practices is provided by the anonymous author of the Quito geographical relation:

> They cultivate … maize in ridges [*camellones*], measuring little more than [*poco más de*] one foot from one to the other. When they wish to sow, they make a hole with their finger and throw in it two grains of maize and a bean; the latter winds around the maize stalks as it grows … The Indians sow potatoes in their maize

---

31 For the abundance of dogs see the passage of Lope de Atienza quoted by Salomon, *Native Lords*, 82.

32 Atienza, "Compendio," page 53, as quoted and translated by Salomon, *Native Lords*, 75–76.

33 Document quoted in Ramón Valarezo, "Del Cacicazgo Andino," 644.

34 The first quote is found in a litigation between caciques of Cayambe and caciques of Tabacundo; see Ramón Valarezo, "Del Cacicazgo Andino," 650. The reference to corrals being cultivated is from 1678 visita to Cayambe, quoted on page 648 of the same article.

> fields, although separate from the maize. They always keep their fields weeded and clean and the soil well cultivated, which helps increase production. Maize is sown such that one clump is about [*como*] a foot from the next.
>
> The tools with which they work the soil are spades [*palas*, possibly footplows] of hard wood, five to six hands long and about one hand wide. In the middle they have a notch which makes a handle for applying force and giving a stronger push. With these they work the soil more easily than with *azadones* ... because they crumble the soil better.[35]

This appears to be a description of the cultivation of maize in wachukuna, much like current practice. Farmers still use their feet to measure the proper distance between planting holes; this "foot" is the length of a foot of an adult male plus adjacent space for the seed holes, about 55 centimeters. The statement that the planting ridges were also "little more than" a foot apart indicates that they were comparable to modern wachukuna,[36] and that the wavelength of the ridging system may have been of the order of 80 centimeters. If two maize seeds of 0.8 gram each were planted in holes 55 x 80 centimeters apart, about 36 kg/ha seed would have been planted, roughly comparable to current practice.

Interplanting, Fallow and Rotation

The anonymous Quito passage above mentions that beans were interplanted with maize. The potatoes in plots next to maize plots suggests potato-maize rotation, either in the same year, or in subsequent years. There is unfortunately very little in the ethnohistorical literature to provide further information on rotations and fallow length. The guess may be hazarded that rotations (in the absence of broad beans, barley, wheat, lentils, and peas) were less important than today.

---

35 Anonymous, "Cibdad," pages 217, 227; translation mine.

36 This passage has been interpreted by both Chantal Caillavet and Frank Salomon as referring to wide raised fields, rather than narrow wachukuna; the phrase "from one to the other" would then refer to maize plants rather than to the ridges. It is true that the word camellón was used at this time to refer to wider raised fields, but it doesn't seem likely that the anonymous author of the geographic relation would refer to raised fields in what was intended to be a generic description of highland agriculture, nor that he would repeat himself by indicating the distance between maize plants twice in the same passage using slightly different language each time ("little more than" a foot, and "about" a foot). Since maize grew in clumps of two, the first reference to a singular noun makes more sense as referring to the ridge rather than to "the (singular) maize." Also, why focus so obsessively on the distance between plants and ignore the much more impressive geometry of the fields themselves? The interpretation of this passage by Caillavet and Salomon is actually even more consistent than my own with my view of the role of highland raised fields as developed in Chapter 7. See Caillavet, "Toponimia," 14, and Salomon, *Native Lords*, 74.

Labor Inputs

Although it has been suggested that native tools were better than the sixteenth century metal azadón[37] it is more likely that non-metal tools were less efficient than metal tools for digging and cutting tasks. In New Guinea, felling trees may take four times longer with stone adzes than with steel axes. Digging ditches may take 2.7 times longer with wooden spades than with steel spades, while grassland reclamation in New Guinea with bamboo knives may take about 2.7 times as much time as it does today in Ecuador with metal tools. It will be assumed for modelling purposes that all tasks except maize harvest and irrigation took 2.7 times as long before the advent of metal tools, a result of averaging the two New Guinea values for grassland reclamation and ditch digging.[38]

Terracing: Northern Highland Air Photograph Analysis

There are apparently no sixteenth century reports of terraces in the Equatorial Andes. We thus are dependent on the discovery of abandoned features which can be interpreted as terraces.

Gondard and López, using air photographs and limited ground checks, have identified 48 "terrace" sites in the northern half of the equatorial Andes

---

37 Anonymous, "Cibdad," 227.

38 For reports on labor time for tree felling and digging ditches in the clay soils of New Guinea, see Axel Steensberg, *New Guinea Gardens: A Study of Husbandry with Parallels in Prehistoric Europe*, New York: Academic Press, 1980, 34–38 and 90. Steensberg found that labor time for digging ditches was 2.75 times more efficient with steel spades. Steensberg also found (page 55) that preparation of fields in grassland required 100 person-days per hectare in New Guinea with bamboo knives; this is 2.7 times as much labor as typically required for grassland reclamation (pudrición) in Ecuador with metal tools (37 person-days per hectare). Shovels were also found to be 2.77 times more efficient than digging sticks at Las Bocas, Sonora, Mexico; see Charles J. Erasmus, "Monument Building: Some Field Experiments," *Southwestern Journal of Anthropology*, 21 (1965), 285. William Denevan used the multiplier of 2.6 to correct for the greater labor involved in prehistoric earth moving, based on Erasmus; see William M. Denevan, "Hydraulic Agriculture in the American Tropics: Forms, Measures, and Recent Research," in Kent Flannery (ed.), *Maya Subsistence*, New York: Academic Press, 1982, 181–203.

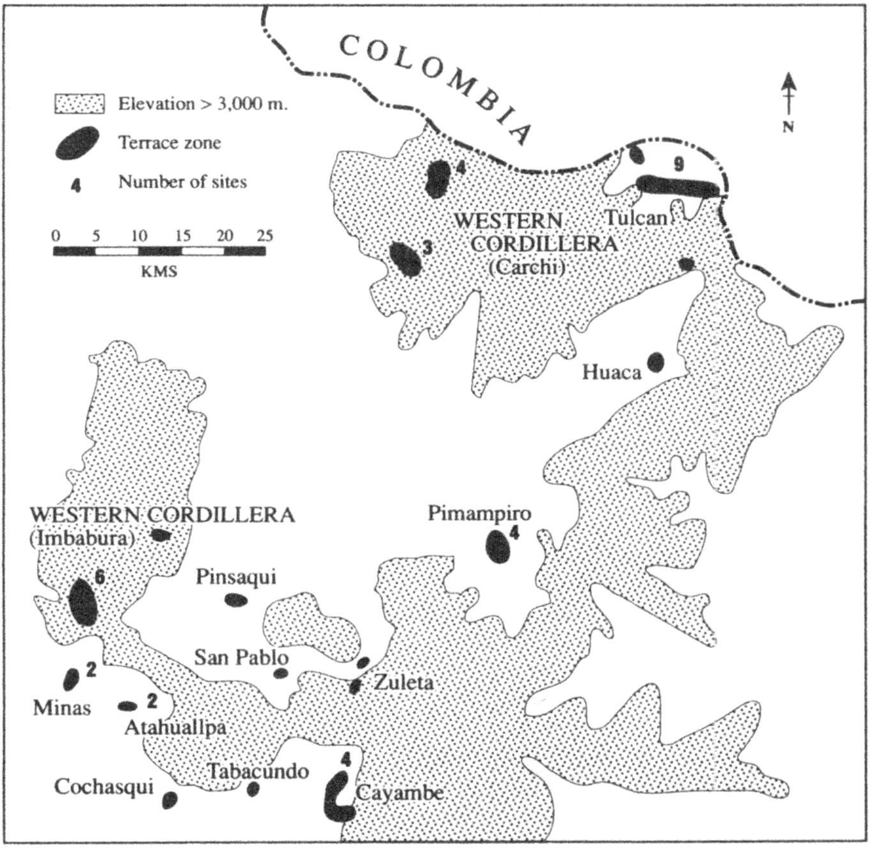

Figure 5.1 Location of "Terrace" Sites, Northern Highlands of Ecuador

(Figure 5.1).[39] The terraces are semi-terraces or sloping terraces, as their surfaces are sloping rather than level and they lack stone retaining walls. These sites can be divided into two categories: (1) Abandoned semi-terraces of the higher elevations, generally well-watered, not currently used for agriculture, and unrelated to archaeologically or ethnohistorically known

39 The following discussion is based upon the maps and measurements presented in Gondard and López, *Inventario Arqueológico*, 135–144. The 47 sites discussed in this chapter of their book appear to inadvertently omit the small (about 10 hectare) site C–135, which should be included. The 48 sites, including the omitted site, contain about 2,627 hectares. Three of the most distant sites of the western flanks of the Andes are omitted from the map; one near Selva Alegre and two near the lower Guayllabamba River.

llajtakuna; and (2) Lower-elevation small terrace sites closely associated with prehistoric llajtakuna. It will be argued that the latter category corresponds with defensive rather than agricultural structures.

Abandoned Semi-Terraces of the High Cordillera

Almost half the measured area of "terracing" is found in 14 sites at high or indeterminate elevations in the western cordillera.[40] Almost all these sites are at high elevations (above 3,000 meters) and most receive over 1,300 mm mean annual precipitation. The only terrace sites in this group which are associated with possible pre-Hispanic settlement features are the southern three sites in the western cordillera of Carchi, which are located close to several *bohío* features (donut-shaped earthworks which were presumably house foundations).

Today, the features of the western cordillera of Imbabura and the northern parts of the Carchi cordillera are under páramo grass, shrub, or low forest vegetation and are neither irrigated nor cropped. The southern set of features of the Carchi cordillera are partly cropped for unirrigated potatoes, broad beans, and small grains.

It would be hazardous to attribute a major agricultural role to these semi-terraces in view of their marginal suitability for agriculture and the lack of association with important settlements.

Lower-Elevation Defensive Terraces of the Llajtakuna

Most of the other semi-terrace sites are located close to ethnohistorically and archaeologically known llajtakuna centers, as indicated by the presence of tolas or similar features. Occurring at lower elevations (many below 3,000 meters), and moderate to dry precipitation values, many of these sites are currently farmed, often with irrigation. It therefore is plausible that these represent pre-Hispanic agricultural terraces. This however was probably not their primary function.

The area of semi-terraces in each site is small, averaging 40 hectares each. The total area of the 34 sites outside the western cordillera is less than 14 square kilometers, a very small proportion of total agricultural land area. If the terracing were primarily agricultural in function, one would have to explain why such a large area of potentially capable of being terraced land was not terraced, and why the terraces do not appear to be constructed in zones of particularly good soils or ease of irrigation.

---

40 The 14 sites in the western cordilleras of Carchi and Imbabura contain 1,245 of the 2,627 total hectares.

A typical set of terraces (one of the five sites near Cayambe)[41] was visited, measured, and sounded with a soil auger. The sloping treads were about 20 meters wide, with risers about 4 meters high. The treads were concave upward, with a minimum slope of 26° (about 50 percent) at their lower ends, increasing to 50° at their upper ends. A series of 1.2 meter deep probes with the soil auger disclosed no buried soil horizons, leading to the conclusion that there never was a flat surface here; these are sloping terraces. Some of the treads were being farmed for unirrigated maize at the time of this study. There were no signs of irrigation or drainage canals, past or present. The topsoil at the bottom of the features shows no signs of accumulation, and depth of topsoil decreases towards the top of the semiterrace treads, indicating that these features are products of erosion into the pre-existing slope surface.

The location and form of the Cayambe terraces strongly suggest house sites located on steep slopes for defensive purposes. This site is located on a commanding hillside position about 4 kilometers from the abandoned tolas and pre-Hispanic agricultural fields of the Cayambe flats. During the Inca wars the people of Cayambe and neighboring areas constructed military strongholds,[42] which very likely involved the location of at least part of their settlements on steep slopes with long parallel walls cut out of the volcanic ash. This interpretation helps explain the presence of abandoned terraces mostly on the steeper slopes but not on gentler slopes; it also helps explain the relatively small area of terraces in the vicinity of pre-Hispanic llajtakuna.

The largest of the four Pimampiro sites was visited and measured.[43] Here, the risers are often more than 3 meters high and the treads are relatively narrow (about 10 meters). The entire slope of these semi-terraces is rich with evidence of dense pre-hispanic settlement, with abundant sherds and kitchen refuse pits.[44] Although this slope is partly under cultivation, it is hardly an attractive agricultural location in comparison with abundant nearby areas of more gently sloping land.

An early geographic relation from the Pimampiro area suggests that these semi-terraces may have been primarily defensive settlement terraces:

> The Indians of this land used to have wars with each other in past times ... In past times, these Indians used to have their houses on some very high hills [*cerros*], for

41 Gondard and López' site P-097, about 4 kilometers from the raised fields and tolas of the Cayambe flats.

42 *Casas fuertes*. See discussion in Larrain Barros, *Cronistas*, Volume 2, 310–314, footnotes 138 and 140.

43 Gondard and López' site I-073. The area of this site as given in Gondard and López is a bit misleading, as about half the area consists not of semi-terraces on steep slopes but rather of semi-circular embankments on gently sloping terrain.

44 José Echeverría assisted with the reconnaissance of this site.

> which they were considered valiant. They used to subjugate the people of the hot valley, because the latter were very lazy and didn't amount to much....[45]

The best analogue for these defensive terraces is perhaps the *cerros de trincheras* (hills with entrenchments) or fortified hill towns of Sonora Province, Mexico. Here walls 2 to 2.5 meters high and 15 to 30 or more meters long were constructed on hill slopes overlooking settled lands. The trincheras were wide enough to accommodate a native house and long enough for a number of them. The surfaces contain abundant sherds.[46]

It is noteworthy that in southern Colombia, agricultural terracing of the temperate elevations of the Andes is unknown but house platforms were cut into the slopes to create settlement terraces.[47] Farther north, large parallel "step-like concentrations of earth about 10 meters wide and up to 100 meters long, the face of each step being 3 meters or more high" with sloping treads and numerous sherds resemble the Ecuadorian settlement terraces.[48]

To conclude, there is little evidence for agronomically significant terracing in the equatorial Andes. A relatively small area of possible prehistoric semi-terracing can be identified in air photographs; and these features are either located in remote and ecologically disadvantaged cordillera sites or are located in close juxtaposition to chiefdom centers as probable defensive settlement features. They probably were farmed, but this was the product rather than the cause of their settlement.

Coca Terraces?

Some of the terrace sites have been interpreted as coca terraces, especially the Pimampiro sites described above.[49] This is unlikely for several reasons.

In Peru, where coca growing is legal, the upper limit of cultivation varies from 1,920 meters in the south to 1,500 meters in the north.[50] It is unlikely that coca cultivation would have taken place much higher than 1,500 meters in Ecuador. The terrace sites of Pimampiro are all higher than

---

45 Borja, "Relación," 251, translated by the author.

46 See Carl Sauer and Donald Brand, "Prehistoric Settlements of Sonora, with Special Reference to Cerros de Trincheras," *University of California Publications in Geography*, 5 (6) (1931), 67–148.

47 Maria Victoria Uribe, "Asentamientos Prehispánicos en el Altiplano de Ipiales, Colombia," *Revista Colombiana de Antropología*, 21 (1977–1978), 155. Also see Gerardo Reichel-Dolmatoff, *Colombia*, New York: Frederick Praeger, 1965, 82, 100.

48 Sylvia Broadbent, "Agricultural Terraces in Chibcha Territory, Colombia," *American Antiquity*, 29 (1964), 501–504.

49 Gondard and López, *Inventario Arqueológico*, 139.

50 See Gade, *Plants*, 176, and Brush, *Mountain*, 74.

this; indeed only two or three small terrace sites have been located below an elevation of 1,800 meters.

There is also good documentation of early coca growing practices, and it is clear that coca fields were located on irrigated flat alluvial land by the Mira River, not on slopes (see section on irrigation, next chapter).

Some Other "Terrace" Sites

The equatorial Andes to the south of Quito has not been surveyed as carefully as the northern highlands, but the general picture is similar. There is no evidence of true stone-walled pre-Hispanic terracing,[51] and although there are some semi-terraces with impressively high and long risers,[52] the agricultural importance of these features is open to question.

Drainage

Borja reported in 1582 that there was a large lagoon near Pimampiro which the Indians drained in pre-Hispanic times to cultivate crops.[53] Gondard and López suggest that this corresponds with a landslide-created depression 1 kilometer southeast of the current site of Pimampiro;[54] the depression has been intensively farmed and today shows no sign of pre-Hispanic drained fields.

Parts of the Cayambe plain are crisscrossed with abandoned ditches about 20 meters apart. These have been interpreted as the remains of "ditched linear fields" constructed in pre-Hispanic times.[55] However, farmers in the area told me that these are modern, constructed for drainage, but have been superseded by more widely-spaced drainage ditches.

Gently sloping flats near Angochagua and at Shanshipamba in Imbabura Province exhibit the remains of what appear to have been drainage ditches about 11 meters apart. The Angochagua features, visible

---

51 In the area called Shanshipamba to the south of Pimampiro and east of Mariano Acosta, a gentle slope exhibits what appear to be semi-terraces separated by crude lines of stones, perhaps associated with settlement. Much more elaborate stone terraces have been found at lower elevations on the east slope of the Andes near Borja; the functioning of these features remains very unclear. See Pedro I. Porras, *Fase Cosanga*, Quito: Ediciones de la Universidad Católica, 1975.

52 Such features are especially abundant in southwestern Chimborazo province. What appear to be prehispanic semiterraces have also been found on the humid eastern slope near Cuyuja to the east of Papallacta; there is evidence that houses were located on these slopes. See Porras, *Fase Cosanga*, figures 20a and 21a.

53 Borja, "Relación.," 248.

54 Gondard and López, *Inventario Arqueológico*, 166.

55 Batchelor, "Camellones."

in aerial photographs, are oriented in such a way as to have channeled water down slope through the flat-floored bottom of a glacial valley; the area is currently in hacienda cattle pastures. The features at Shanshipamba became visible when brush was removed from this alluvial terrace site after World War II.[56]

Even more intriguing are the widely spaced (10 to 30 meters apart) lines visible on aerial photographs on steeper slopes south of Pimampiro, first reported by David Preston as "field lines,"[57] and later studied in more detail by Preston and the author.[58]

These features occur on relatively clay-rich soils (over 30 percent clay content)[59] and in many cases have been partially filled with slope wash. Many if not all the field lines appear to precede the current agricultural reclamation of this district.

Colimburo

At Colimburo, one of the field line sites, the dark topsoil is relatively deep (30–75 cm) by Ecuadorian slope standards, suggesting that these relatively heavy soils are not readily eroded. The light yellowish brown loessal ash subsoil is difficult to work, but is not cemented into a hardpan. Subsurface mottling is sometimes present suggesting perched water tables. During wet weather, the topsoil becomes difficult to work and slippery for vehicle or foot transportation.

At Colimburo, the field lines on 40 percent slopes currently correspond with depressions 0.5–1.0 meters deep. A road cut across a set of field lines indicated that they correspond with partially filled artificial ditches 1.5 meters deep, having a trapezoidal cross section, narrowing from 1.6 meters at the top to 0.5 meters at the bottom. When interviewed, farmers at Colimburo said the field lines were field boundary markers (*linderos*) and that they were also useful for directing runoff from downpours (*crecientes de agua*), in combination with smaller ditches running diagonally across the slope. Farmers denied that the lines performed any irrigation or soil fertility management function, or that specialized planting in depressions and intervening surfaces was necessary or desirable. Much of the area is currently farmed for barley and wheat.

The author and David Preston now interpret these field lines as the remains of pre-Hispanic ditched fields, which functioned primarily to facilitate speedy runoff from areas of heavy soils, enabling easier

---

56 Based on interview with the hacienda administrator, 1984.

57 See David A. Preston, *Field Lines in Northern Highland Ecuador: Preliminary Account of Field Observations*, Leeds: University of Leeds School of Geography, Working Paper 380, Revised Version, 1984.

58 Knapp and Preston, "Evidence."

59 PRONAREG-ORSTOM, *Mapas de Suelos*, Pimampiro and Mariano Acosta Sheets.

cultivation, reducing waterlogging, and limiting landslide hazard. An important secondary function could have been field bounding; it is possible that there was some crop differentiation near the ditches.

The basic pattern of the ditches is artificial, although natural factors could certainly have helped in ditch deepening. An incremental process of construction aided by natural erosion would have involved relatively little labor.[60] Similar ditched fields have been found on slopes in Colombia, and similar functions and ages have been suggested.[61]

In the following chapters, the relatively more important prehistoric technologies, irrigation and raised fields, will be discussed.

---

60 In this landscape there are both natural gullies and gullies which have subsequently been filled by slope wash. The latter, as seen on the road between Pimampiro and Mariano Acosta, have wavelengths of 45 meters and originally were 5 meters deep. The scale of these features is easily distinguishable from that of the field lines. There are no naturally eroded slopes in Ecuador which resemble the spacing and patterning of the relic field lines.

61 See Warwick Bray, Leonor Herrera and Marianne Schrimpff, "Report on the 1981 Field Season in Calima," *Pro Calima (Basel)*, 3 (1983), 2–31; and Robert Eidt, *Advances in Abandoned Settlement Analysis: Application to Prehistoric Anthrosols in Colombia, South America*, Milwaukee: University of Wisconsin at Milwaukee, The Center for Latin America, 1984. Drainage and landslide control have been suggested as functions of modern vertical ditches at Espinar, Peru; see Benjamin S. Orlove, "Integration through Production: The Use of Zonation in Espinar," *American Ethnologist*, 4 (1977), 96.

Chapter 6

PREHISTORIC IRRIGATION

Most agricultural intensification in pre-Hispanic times took the form of irrigation or raised field systems. Both increased production, but in different niches and with different technologies, and with different efficiencies in terms of land and labor. Irrigation persists in the Ecuadorian highlands while raised fields are essentially extinct.

It is extraordinarily difficult to date such agricultural infrastructural features as raised fields and irrigation canals.[1] Ceramics or datable carbon material usually occur in contexts of disturbance and redistribution of sedimentary layers, so that they can not readily be related to the agricultural feature. In addition, with a few exceptions, the observations of Spaniards and the archival documents of the sixteenth century ignore the specifics of agricultural technology. As a result, it is necessary to assemble a wide range of disparate evidence to demonstrate the plausible antiquity of raised fields and irrigation. This chapter will provide evidence on irrigation and the next chapter will provide evidence on raised fields, as a preparation and justification for the modelling of pre-Hispanic land use in Chapter VIII.

Irrigation: General Comments

It is commonly believed that the irrigation systems of the Ecuadorian highlands are relatively new, having been constructed in the last 150 years.[2] Relatively little work has been devoted to highland Ecuadorian irrigation

---

1 But see Erickson, *Archaeological Investigation*, 188, where in the absence of datable radiocarbon samples, 7 thermoluminescence dates were obtained from raised field contexts, and the discussion of my raised field radiocarbon sample in the next chapter.

2 D.C. Anderson, writing in Appendix D, Keller et al., *Ecuador Irrigation*.

history, as compared with the history of the Peruvian coastal oases or, more recently, work on the irrigation of highland Peru.[3]

The characteristics of traditional irrigation have been discussed in Chapter 3. Many of these characteristics can be traced to Spanish contributions. In the canals themselves, the use of a graduated cylindrical hole in stone to control water volume (*óvalo*) is a Mediterranean introduction. Gristmills for wheat, woollen mills (*obrajes*), and sugarcane mills (*trapiches*) were established on preexisting canals or on new canals specially constructed for the purpose. The major innovations of the Spaniards, however, were in the crops themselves and the institutions of land and labor control and marketing which organized and deployed production. Within a short time after the conquest, the Spaniards had introduced grapes, olives, figs, apples, quince, peaches, oranges, limes, lemons, pomegranates, citron, sugar cane, bananas, plantains, melons, eggplants, garbanzo beans, cabbage, lettuce, radish, and mustard, and were irrigating these products in various highland oases in Ecuador.[4]

Much of the organization of irrigation production was in the hands of religious orders, which located potential canals, supervised their maintenance, controlled the appropriate lands, and managed the Indian or black slave labor. At somewhat higher elevations, irrigation played a role in the maintenance of pastures (including common pasture or ejido lands) and the production of grain crops both within religious or lay haciendas and on the small plots of indigenous producers.

Given this strong Spanish overlay, the task of determining the characteristics of pre-Hispanic irrigation requires a combination of archaeological, ethnohistorical, and geographical detective work. A number of valleys in the northern highlands of Ecuador were surveyed by the author between 1984 and 1987 for traditional canals and pre-Hispanic settlements (tolas); this information combined with material from early accounts and archives, permits some reconstruction of the scale and nature of pre-Hispanic irrigation. The sites surveyed for irrigation are displayed on Figure 6.1, together with isohyets of precipitation.

In this area there is good evidence for two kinds of pre-Hispanic irrigation: (1) irrigation for maize, at elevations above 2,300 meters; and (2) irrigation for cotton, fruit trees, and sometimes coca, below 2,300 and usually below 2,000 meters.

---

3 William Mitchell, "Irrigation and Community in the Central Peruvian Highlands," *American Anthropologist*, 78 (1976), 25–44; Guillet, "Terracing;" Sherbondy, *Canal Systems*.

4 See Borja, "Relación," 250; Garcés, *Libro de Proveimientos*, page 101; Aguilar, "Relación," 264; Paz Ponce de León, "Relación," 238.

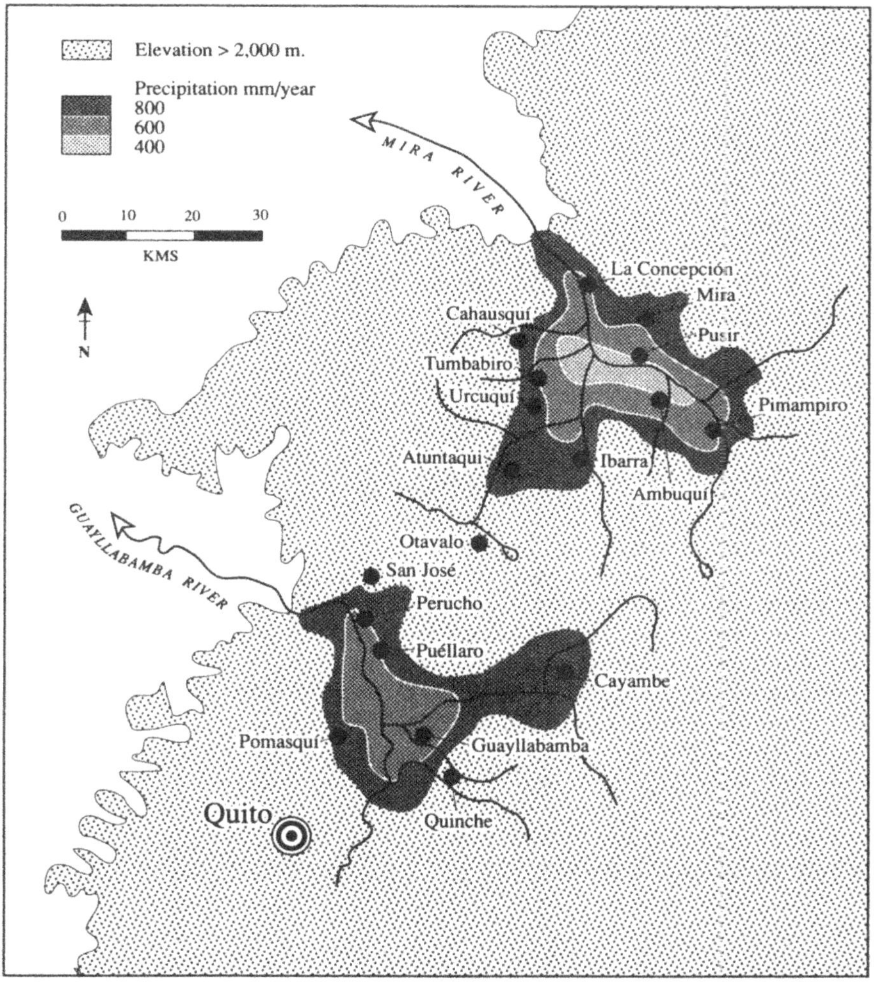

Figure 6.1 Precipitation Map giving Location of Irrigation Study Sites

Irrigation for Maize

Evidence for pre-Hispanic maize irrigation was found in the valleys of Pimampiro, Atuntaqui, El Quinche, and Pomasquí (Figure 6.1) as well as in Chambo and San Andrés farther south (Figure 1.2). The best evidence for irrigation which might not only be pre-Hispanic but also pre-Inca is from the area of Pimampiro.

PIMAMPIRO

Today, the gentle slopes around the town of Pimampiro (Figure 6.1) are irrigated with waters of the Acequia del Pueblo de Pimampiro; the major crops include beans, avocado, tomato, chili peppers, anise, corn, peas, sugar cane, and fruits.

The Acequia del Pueblo is 20 kilometers long and begins at an offtake on the Chamachán River at 3,030 meters elevation. The offtake until recently was a "rustic" structure of stones and boulders. The earthen canal flows along mountain slopes and through 17 tunnels excavated in the underlying volcanic ash, delivering about 170 liters/second of water to an irrigated area of 600 hectares with 414 landowners. Maps of this canal are available elsewhere.[5]

The area of Pimampiro in pre-Hispanic times was part of the Chapí chiefdom. There were in the 1580s two Chapí groups: the woodsmen of the south, and the northern Chapí near Pimampiro. The southern Chapí spoke two languages: the language of Otavalo and another similar to that of the Quijos of the Amazon basin. The northern Chapí spoke only the language of Otavalo. Both groups lived on high slopes and dominated the peoples of the tropical Mira valley to the north.[6]

Archaeology helps locate the sites of these two Chapí groups. Two tola sites have survived into modern times: one at Paragachi (now destroyed), 2 kilometers north of Pimampiro, and the other at Casachupa, 9 kilometers southwest of Pimampiro. The latter is located adjacent to abundant settlement terraces with pre-Incaic ceramics. This may have been the most important center of the Chapí woodsmen, although other large sites exist at Shanshipamba, 11 kilometers south of Pimampiro, and Pueblo Nuevo, 8 kilometers south of Pimampiro. None of these sites have irrigation today.

The northern half of Chapí was already relocated in the new town of Pimampiro in the 1580s. The original settlement was probably located at La Mesa, 4 kilometers southeast of Pimampiro. The settlement terraces of this site have been described in the preceding chapter. The best agricultural land

---

5 For maps and detailed descriptions of the contemporary characteristics of this canal, see Mothes, *Pimampiro's Canal*, and Mothes, "Acequia." An archaeological map of the vicinity was published in Knapp and Preston, "Evidence."

6 See Borja, "Relación."

would have been on the flats south of Pimampiro to the west, which would have been reached by cabuya fiber bridges. These flats were irrigated by a long canal:

> The water which these Indians of the town of Pimampiro use is from a ravine which is in the aforementioned woodlands of Chapí, and they bring it by a canal more than two leagues to this town; and with the canal water they irrigate the maize crops which they have.[7]

The indicated location and length of this canal permits identifying it with the modern Acequia del Pueblo.[8] It is worth noting that the area near the offtake of this canal has an abundant surface scatter of pre-Inca ceramics, and had an apparently precolumbian stone stela with zigzag and spiral motifs.[9]

A few years later, the canal of Pimampiro was apparently briefly abandoned. Ordóñez de Ceballos wrote in 1614:

> That town [Pimampiro] used to have deficient water, and many fields were lost ... we discovered some canals from the times of the Inca and we saw how a great quantity of water could come ... and so many Indians came to help, that in 15 days they built a canal five leagues long and a vara and a half deep and a vara wide, which carried two bueyes of water...[10]

This also appears to refer to the Acequia del Pueblo, with some boastful exaggerations as to size and novelty. What appears to be described is not the construction of a canal but rather a cleaning and repair operation on a canal which had fallen into disuse. The attribution to the Incas is not to be taken literally; there is no evidence of any significant Inca presence in this area.[11]

Apparently, the pre-Inca Acequia del Pueblo remained in use through the Inca conquest and into the 1580s, the only major changes being the Spanish relocation of the houses of the cultivators from the hill slopes to a new settlement (reducción) in the midst of the fields. It is also clear that the

---

7 Borja, "Relación," 249.

8 The reference to the Chapí woodland allows us to locate the canal's origin in the Casachupa-Shanshipamba area. Its origin in a nameless ravine rules out either the Mira or the Pisque as possible water sources. The distance is also appropriate: Borja's usage of the league corresponds to a measure of about 5 kilometers. (The distance from Pimampiro to Quito is given by Borja as 18 leagues, while the actual linear distance is 93 kilometers). The linear distance from the Acequia del Pueblo offtake to the town of Pimampiro is 13 kilometers.

9 This stela, like many other monuments of the vicinity, was brought to the bishop's palace of Ibarra by Silvio Luis Haro.

10 Pedro Ordóñez de Cevallos, "Historia y Viaje del Mundo del Padre Agradecido [1614]," *Cronistas Coloniales* 2, Puebla, Mexico: Biblioteca Ecuatoriana Mínima, Editorial Cajica Jr., 1960, 475–519.

11 See the discussion in Salomon, *Native Lords*, 188–192.

major crop was maize, with a shift into such commercial crops as wine grapes.[12]

Today the Acequia del Pueblo irrigates 600 hectares. It receives supplementary water from the La Chorrera offtake, which was apparently constructed in late colonial times. If the water from this source is subtracted, the canal would still have been able to irrigate 400 hectares.[13] The irrigated area has an elevation from 2,700 meters to under 2,000 meters, and an average annual precipitation (Pimampiro station) of about 600 millimeters.[14] As will be seen, the area cultivated, elevation, and precipitation are not much different than other plausibly pre-Hispanic areas of maize irrigation.

OTAVALO-ATUNTAQUI

A short distance to the south of Otavalo (Figures 1.2 and 6.1), the San Pablo Lake is drained by the Jatunyacu River, which passes over the Peguche waterfall and contributes to the Ambi River. A number of canals begin at or near the waterfall, but the oldest and most important was the Old Peguche Canal.[15] This canal has been modified in its lower course so as to supply water to a hacienda area, but the original lower course can still be determined, since the original lower course continues in use with a different water supply.

The Old Peguche Canal begins at a rustic offtake (formed primarily of boulders) immediately below the Peguche waterfall near Otavalo. Here the canal is now called the Acequia de Pérez.[16] The earthen canal flows northward along the riverside flats, and passes the ruins of the colonial Peguche woollen mill (which this canal once supplied). It then passes through what are now flat pastures near Quinchuquí.

---

12 According to one source, this area was depopulated about 1679 as wine grapes went out of production and all the Indian cultivators fled to the eastern slopes of the Andes, so that an ethnically distinct population took over the irrigated area. For this story, see Juan de Velasco, *Historia del Reino de Quito en la América Meridional [1789]*, Quito: Editorial Casa de la Cultura Ecuatoriana (3 Volumes), 1977–1979, Volume 3, 105–106.

13 The Chorrera offtake supplies perhaps one-third of the water flowing in the Acequia del Pueblo. See Mothes, *Pimampiro's Canal*, 114–119.

14 PRONAREG-ORSTOM, *Estudio Hidro-Meteorológico*, 54; the exact value is 588 millimeters for the 1964–1973 period.

15 The term "Old Peguche Canal" is not in local use but seems convenient to describe this dismembered canal. The upper portion is called the Acequia de Pérez, and the lower portion is called the Acequia de San Antonio. For a map of the canals between Otavalo and Atuntaqui, see Gregory Knapp, "Riego Precolonial en la Sierra Norte," *Ecuador Debate (Quito)*, 14 (1987), 38.

16 12mmediately below it, another canal, also called by some Acequia de Peguche, irrigated the lands of the Peguche Hacienda; it is now abandoned.

At this point, the Old Peguche canal originally continued to Atuntaqui (Figure 6.1), but has been rerouted to deliver water to Hacienda Agualongo just to the south of Atuntaqui. This appears to have been the solution to a water dispute of the mid-nineteenth century.[17]

The original lower route of the old Peguche canal is today supplied by water from a new canal beginning above the Peguche waterfall.[18] The lower canal passes by large tola sites (Pinsaquí and Agualongo) before dividing into three branches at a division structure (óvalo). One branch travels below the giant tolas above Atuntaqui, drops down a ravine, and continues by other tolas to the settlement of Chaltura and the low (under 2,200 meters) tola-strewn flats of Santiago. Another branch supplies water to the area around the giant tolas of Atuntaqui. A third branch, called the "Peguche" branch, travels to another distribution box (Los Ovalos) by a tola, from where further branches provide water to Chaltura, Natabuela, Bellavista, and San Antonio de Ibarra.

In all, the old Peguche canal now has a flow of 563 liters/second at its offtake, but much of this is lost through evaporation and infiltration lower in its route. In the late 1970s, this canal and its new lower course irrigated 470 cultivated hectares, primarily corn and beans.[19]

There is evidence that the area embracing the San Pablo lake and extending down to Atuntaqui and beyond was under the mandate of a single paramount lord in pre-Hispanic times. A chief named Otavalo was captured by the Spanish in 1534; the political center of Otavalo named for this lord was located next to San Pablo lake.[20] About 1577, there were 37 parcialidades whose principales were subordinate to Otavalo; it appears that while the top of the political hierarchy was centered around San Pablo Lake,[21] another major center of power was at Atuntaqui, where the ethnic lord controlled the largest population of the region and whose very name (Apoango) suggested special importance. "Apo" is an Inca prefix

---

17 A document of 1858 refers to the claim that José Pérez Calisto, owner of hacienda Agualongo, had purchased water from the canal going from Peguche to San Antonio, but had arbitrarily appropriated almost all the waters of the canal. At this time there was clearly a single canal supplying Peguche, Quinchuquí, Agualongo, San Antonio, and Atuntaqui. See litigation between Señor José María Pérez Calisto and Señor Jijón concerning "la acequia sacada de los citios de sus haciendas de Peguchi i Quinchuqui," (1858), Archivo Histórico del Banco Central, Ibarra, Juicios, Paquete 259 (1846–1858).

18 Now called the "San Antonio" canal.

19 "Inventario de los Usos de Agua, Subcuenca del Rio Ambi," (1977), Instituto Ecuatoriano de Recursos Hidráulicos División de Recursos Hidrológicos, Quito.

20 Chantal Caillavet, "Etnohistoria Ecuatoriana: Nuevos Datos Sobre el Otavalo Prehispánico," *Cultura (Quito)*, 5(11) (1981), 109–127.

21 Including the areas of San Pablo, Otavalo, and San Juan Imbaquí.

corresponding to high rank, and the lord of Atuntaqui was the only one of the region to bear this title.[22]

As late as 1609, the area at Peguche (Piguchí) was one of the most prized possessions of the paramount ethnic lord of Otavalo, but by 1616 the area had been sold to a Spaniard to allow construction of the Peguche obraje.[23] Thus at least the upper portion of the Old Peguche Canal can be assumed to have been functioning by 1616, if not earlier.

It is unfortunately difficult to trace the colonial irrigation history of the Atuntaqui area, in part because the major haciendas (including Agualongo and the Santiago valley below Chaltura) were in the hands of the Jesuits,[24] and it has proved difficult to find the archival records of the Jesuit haciendas of the northern Ecuadorian highlands. Also, since the town of Atuntaqui lies considerably below the old Peguche Canal, its own documentation emphasizes local canals derived from local springs.[25]

The Old Peguche canal is referred to in an 1858 document as "the canal brought from the locations of his [Señor Jijón's] haciendas of Peguchi and Quinchuquí;" Jijón was at this time "obligated to turn over to the citizens of the towns of Atuntaqui and San Antonio a quantity of water," and was also selling water to the Hacienda Agualongo. There is no reference to construction of the canal in this document, which suggests the canal had considerable antiquity.[26]

To recapitulate, the evidence for the antiquity of the old Peguche canal includes (1) its traditional construction and rustic offtake; (2) abundant pre-Hispanic settlement remains and ceramic scatter along its entire course and especially in the areas of its destination (Atuntaqui, Chaltura,

---

22 Caillavet, "Estructura Básica;" and Caillavet, "Adaptación de la Dominación," especially page 415.

23 See Chantal Caillavet, "Caciques de Otavalo en el Siglo XVI: Don Alonso Maldonado y su Esposa," *Micelánea Antropológica Ecuatoriana*, Year 2 (Number 2) (1982), especially the transcription of Don Alonso Maldonado's testament (1726) on pages 45–51; also see Caillavet's footnote 2, page 53. The Peguche houses and fields are mentioned in the will immediately after the chief's houses in Otavalo and Quito, and before any other lands, and are bequeathed to his legitimate son and political heir Don Pedro Maldonado.

24 Victor Alejandro Jaramillo, *Corregidores de Otavalo*, Otavalo: Instituto Otavaleño de Antropología, 1972, 139.

25 In a very early document, in 1616, Father Juan Cortés asked for a spring below the royal road to irrigate some lands there; see Jorge Garcés (ed.), *Libro Primero de la Villa de San Miguel de Ibarra*, Quito: Talleres Tipográficos Municipales, Publicaciones del Archivo Municipal Vol. XV, 1937, 557. In one document, the spring-derived Acequia Alta and Acequia Baja are shown following their modern routes; see map on folio 22 of incomplete document, (1860?), Archivo Histórico del Banco Central, Ibarra, Juicios, Paquete 209 (1860–1865). The original name of Antonio Ante was Acequia Alta, suggesting the importance of this canal to its modern development; see Zumárraga, *Monografía*, 199.

26 "Revindicación de las aguas," AHBC/I.

Natabuela), areas which would have been vulnerable to drought without irrigation; (3) evidence of high population density and political importance in the destination area in the sixteenth century; (4) evidence of importance of its offtake area (Peguche) as the favored lands of the highest chief of Otavalo; (5) its use to supply water to a colonial mill, constructed by 1616; (6) evidence of its existence in the early nineteenth century, with no mention of recent construction; (7) strong traditional claims by commun-ities rather than haciendas to its water; and (8) the lack of other canals in the vicinity of equivalent claims to antiquity.[27]

The potentially irrigated land around and below Atuntaqui consists both of flats and gently sloping alluvial terraces, extending from about 2,500 meters at the Atuntaqui tola site down to about 2,300 meters in the vicinity of Chaltura and San Antonio de Ibarra. Precipitation declines from about 770 millimeters at Atuntaqui; there are no meteorological stations at Chaltura or Natabuela, but precipitation can be interpolated as about 700 millimeters.[28] The irrigated area, precipitation, and elevation of the Old Peguche Canal are similar to those of the Acequia del Pueblo de Pimampiro.

QUINCHE

The Acequia del Pueblo del Quinche begins in the Quinche River, a little above the town of the same name, and near perhaps the southernmost tolas in the highlands (Figure 6.1 and part of Figure 6.2). In 1974 the offtake was still rustic and the canal irrigated 1,800 hectares.[29] Much of the irrigated area was in the vicinity of Ascázubi, which has acted to take control of the water in its vicinity (Figure 6.2). Previously, the canal continued past two tolas below Hacienda Guanguilquí (and the larger group of tolas to the east of Guanguilquí) to irrigate lands around Hacienda La Merced. This stretch would also have been ideally situated to irrigate the area around Mesa Tola and the vicinity of the modern town of Guayllabamba.

A document of 1577 refers to a pueblo "of Quinche and Guayllabamba," suggesting the political unity of the Quinche-Guayllabamba

---

27 The "La Victoria" canal was constructed in the early nineteenth century. See Zumárraga, *Monografía*, 219–223; "Inventario Subcuenca del Rio Ambi," INERHI.

28 PRONAREG-ORSTOM, *Estudio Hidro-Meteorológico*; PRONAREG-ORSTOM (Programa Nacional de Regionalización Agraria - Office de la Recherche Scientifique et Technique Outre-Mer), *Isoyetas Medias Anuales e Histogramas Mensuales, Período 1964–1973*, Quito: Instituto Geográfico Militar (1:1,000,000 Map), 1979.

29 See INERHI (Instituto Nacional de Estudios de Recursos Hidráulicos), *Inventario de los Usos de Agua en las Cuencas de los Ríos Chiche, Guambi, Uravia, y Coyago, Proyecto Esmeraldas - Unidad Técnica*, Quito, 1974.

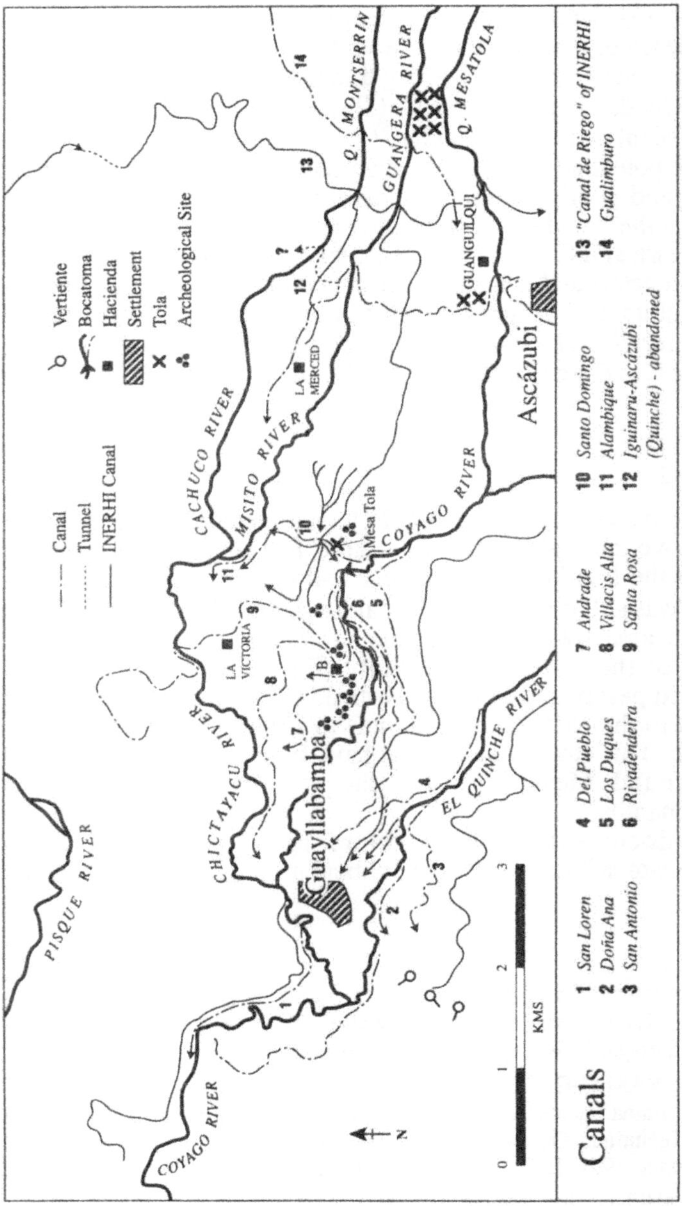

Figure 6.2 El Quinche and Guayllabamba Irrigation Zones

region,[30] uniting the tola sites of Mesatola, Guanguilquí, and El Quinche (Figure 6.2).

There is early evidence of irrigation in this region. About 1558, the Mercedarian Order received lands "near Guayllabamba;" a local ethnic lord named Don Juan Puratico contested this grant:

> [Don Juan Puratico deposed that] the aforesaid land belonged to an Indian named Cachequinche grandfather of the aforesaid Don Juan Puratico and that Topinga Yapanque took it from him and sowed dry season [*michiai* = *michka*?] maize in it and that later, in the time of Guancaba and Atabalipa they used to do the same in the aforesaid lands and after the Spaniards entered they weren't cultivated since the canals were ruptured and it would be difficult to open them again. And [he deposed that] he loved the lands very much since they belonged to his grandfathers ... Asked if he wanted to cultivate them to his liking, he said no: he had lands higher up, half a league from there, and that his pueblo was three leagues from there...[31]

These lands, given to La Merced, must have been above the present-day Hacienda La Merced, in the area of the old Hacienda Monteserrín.[32] The canal could not have been any other than the Acequia del Pueblo del Quinche; the lands the chief still controlled "half a league" upslope suggest the area of Guanguilquí.

The reference to the Inca indicates the possibility that maize irrigation in this zone was initiated by them. Indeed, in 1577 there is a reference to a "Don Juan Mitima," as chief of the region,[33] perhaps the same individual as Don Juan Puratico; the word "mitima" suggests association with an Inca colony (mitmac), while Puratico was a place "next to" Quinche.[34]. By 1582, the town of Guayllabamba was shared by a priest with another town called "Elguanca," the latter probably can be referred both to the toponym El Guanca near El Quinche and with the Guangas or Wanka ethnic group of the Jauxa region of Peru brought into this area.[35]

Although this might seem conclusive evidence that the canal at El Quinche is Incaic rather than pre-Incaic, it is worth recalling Caillavet's suggestion[36] that the mitmajkuna in the zone of Quito were primarily soldiers, who came without their families and rapidly integrated into the

30 See source cited in Caillavet, "Adaptación de la Dominación," 408.

31 Monroy, "Convento," 159.

32 The oldest topographic map of the region shows Hacienda Monteserin just above the Acequia del Pueblo, between the Río Guangera and Quebrada Montserrín (Río Cachuco). A hill near the present Hacienda Merced is also called Monteserin on this map. Monteserin (Montserrín) is a toponym associated exclusively with Mercedarians. There are no Mercedarian toponyms in the area to the west, near Guayllabamba. See SGN (Servicio Geográfico Militar), *Guaillabamba, (Provincia de Pichincha)*, Quito: Servicio Geográfico Militar (1:25,000 Map), [1940?].

33 Caillavet, "Adaptación de la Dominación," 408.

34 Salomon, *Native Lords*, 164.

35 Salomon, *Native Lords*, 163.

36 Caillavet, "Adaptación de la Dominación," 403–423.

surrounding society, rather than "economic" mitmajkuna. The tolas at El Quinche, Guanguilquí, and Mesatola are of authentic local design and much of the ceramics are indistinguishable from those of other tola sites.

The area which was probably irrigated below the canal near Hacienda La Merced is a gently sloping terrace of stratified volcanic ash bounded on its lower edge by a fault scarp. The terrace area is one kilometer wide and several kilometers long, with elevations varying from 2,500 meters at the lower margin to 2,600 meters; mean annual precipitation may decrease from 800 millimeters at the upper end of this tract to 600 millimeters at the lower end.[37]

CHAMBO AND SAN ANDRÉS

In Chimborazo province, the Inca had maize fields "past the Chambo bridge," in an area which today is the irrigated maize and vegetable producing district of Chambo (Figure 1.2). Although Inca irrigation is not explicitly mentioned, it can be provisionally assumed given the fact that the Inca gave the local ethnic lord compensatory and probably irrigated lands in the area of Píllaro far to the north. This ethnic lord also appears to have been the paramount lord for the region, at least after the Inca conquest [38]

The probably irrigated area of Chambo would have corresponded then as it did in 1938 and does now to the gently sloping volcanic terrace between 2,700 and 3,000 meters elevation, with approximately 600 millimeters mean annual precipitation; a few square kilometers are involved.[39]

There is also a brief mention of early indigenous canals at San Andrés, to the northwest of Riobamba (Figure 1.2), not taken from the nearby river but rather brought from the heights of the mountains into the area. Maize was a principal crop, with the fields and houses being surrounded with *cabuya* hedges.[40] San Andrés is about 3,000 meters in elevation and commands the Guano River valley, which descends to 2,650 meters in the 8 kilometer stretch to the east. Mean annual precipitation here is about 600 millimeters.

---

37 Based on interpolation from PRONAREG-ORSTOM, *Isoyetas*.

38 This was the ethnic lord of Guayllabamba. Salomon, *Señores Etnicos*, 194–195.

39 See Gustavo Antonini et al., "Chambo: Evolución del Paisaje y su Relación con el Potencial Productivo Agrícola," *Revista del Centro Panamericano de Estudios e Investigiones Geográficas (Quito)*, 7 (1981).

40 Fr. Juan Paz Maldonado, "Relación del Pueblo de Sant-Andrés Xunxi para el muy ilustre Señor Licensiado Francisco de Auncibay, del Consejo de su Majestad y su Oydor en la Real Audiencia de Quito [1582]," in Marcos Jiménez de la Espada (ed.), *Relaciones Geográficas de Indias - Perú, Volume 2*, Madrid: Biblioteca de Autores Españoles 184, Ediciones Atlas, 1965, 261–262.

POMASQUI

North of Quito, there is a large gently sloping plain at about 2400 meters elevation, extending from Pomasquí to Hacienda Oasis (Figures 6.1, 6.3). A number of canals are derived from the Cotocollao River to enable irrigation of about 235 hectares on both sides of the river.[41]

These canals begin at a place with the toponym "Las Tolas," although unfortunately today no tolas survive in the landscape.[42] A little to the north of the town of Pomasquí, a natural hill has a heavy dispersion of pre-Inca ceramics.

Immediately after the Spanish conquest, in 1535, the Pomasquí valley was distributed as gardens to the citizens of Quito. A document detailing this distribution has survived; two canals are mentioned, a high canal to the west and a principal canal to the east. The document appears to relate the irrigation to Cañari Indians who were living close to Pomasquí. The details of distribution of individual plots indicates that irrigation extended 3.7 kilometers to the north on the eastern side of the river, and 2.2 kilometers to the north on the western side of the river. About 2.5 square kilometers would have been irrigated,[43] in a region with about 600–700 millimeters mean annual rainfall.[44]

A document from 1573 indicates that there were no cultivators of this plain prior to the Inca arrival, and that it was settled with mitmajkuna, presumably as the irrigation system was developed. There are indications that these colonists, unlike those at El Quinche, came from nearby, including the dry valley areas around the Guayllabamba River; some of them at least would have been familiar with pre-Incaic irrigation.[45]

---

41 INERHI (Instituto Ecuatoriano de Recursos Hidráulicos), *Inventario de los Usos de Agua en la Cuenca del Río Monjas: Cuadros Catastrales y Mapas de Ubicación, Proyecto Esmeraldas - Unidad Técnica*, Quito, 1973.

42 Plaza Schuller identified possible tolas in the flats to the north of Pomasquí using air photographs; these could not be found in the field. See Plaza Schuller, *Contribución*.

43 See Jorge Garcés (ed.), *Libro Primero de Cabildos de Quito [1529–1538]*, Quito: Archivo Municipal de Quito, Publicaciones No. 1, 1934, 108–115. The lands were distributed according to a measure called a cuerda, equivalent to the width of the Quito plaza (about 90 meters); the lands on the east side extended for 41 cuerdas; the lands on the west side extended for 21 cuerdas plus two additional plots. If the additional plots were 2 cuerdas wide each, then the lengths would have been 25 x 90 = 2,250 meters and 41 x 90 = 3,690 meters.

44 PRONAREG-ORSTOM, *Isoyetas*.

45 See the document quoted and discussed in Salomon, *Native Lords*, 164–167.

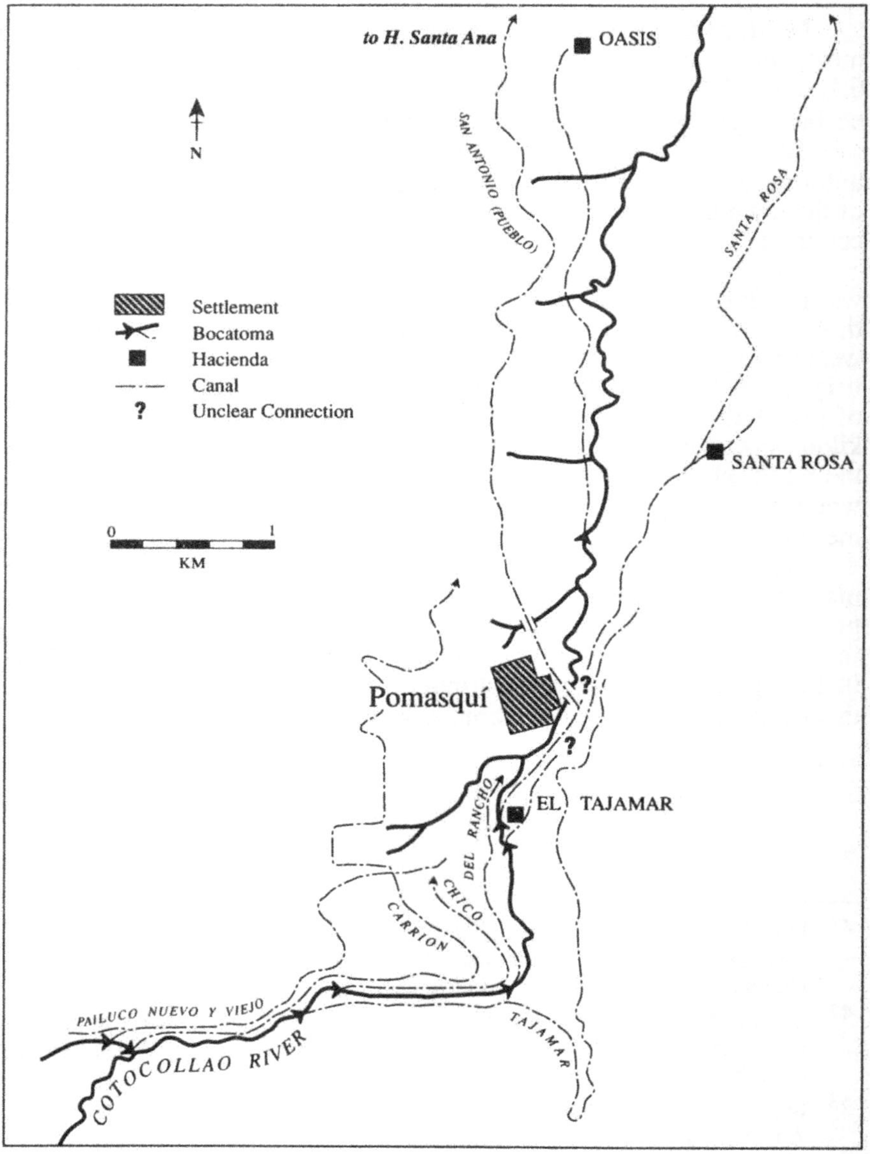

Figure 6.3 Pomasquí Irrigation Zone

Summary: Maize Irrigation

The cases detailed above suggest several common characteristics. All the sites are located in ecologically similar areas: gently sloping volcanic ash terraces with deep soils and about 600–700 millimeters per year precipitation. Elevations vary from 2,300 to 3,000 meters.[46] The relatively high precipitation would have allowed greater efficiency in the use of irrigation water. In all the sites discussed so far, maize was the principal crop. And most of the sites involved the irrigation of several (2.5 to 5) square kilometers, suggestive of a certain optimal scale for such irrigation projects.

It is also noteworthy that the irrigated areas were usually controlled by the same chiefdom that controlled the water sources, often under an arrangement where ethnic lords of the irrigated area and the irrigation source were almost equally powerful. Thus the Chapí woodsmen controlled the Acequia del Pueblo's sources, while the lower Chapí group occupied the area under irrigation. The dual "pueblo" of Quinche and Guayllabamba would have controlled both areas there. And the political union of Otavalo and Atuntaqui (under Hernando Apoango) also fits this general pattern.

Irrigation for Tropical Specialty Crops

The greater part of the probably pre-Hispanic irrigation systems were not devoted to production of maize for subsistence but rather for the production of subtropical specialty crops. These sites will be discussed so as to further indicate the evidence for pre-Hispanic and especially pre-Incaic irrigation.

CAHUASQUI

In 1582, there were several hundred Indians living in Cahuasquí (Figure 6.1); in a hot valley one league to the east they cultivated cotton, guayabas, guabas, pineapples, passion fruit, and pepino, as well as the subsistence crops of corn, beans, manioc, sweet potatoes, and peanuts.[47] This hot valley can be identified with the area of La Victoria (Puchimbuela), 5 kilometers east of Cahuasquí, where there is a gently sloping irrigated land between 1,600 and 2,000 meters.

Coca was also grown in this area, at Palacara, 7 kilometers northeast of Cahuasquí; by 1639 the cacique of Salinas controlled at least part of this

46 Only the lower stretches of the Pimampiro irrigated area go below 2,300 meters, to about 2,000 meters. The sixteenth century documents nevertheless do not mention cotton or other tropical crops for the Pimampiro irrigated lands.

47 Aguilar, "Relación," 245–246.

coca land.[48] Indians of Lita on the western slope of the Andes were also settled here.[49]

By 1616 there is explicit mention of canals here,[50] but the dry climate would have made the earlier cultivation of cotton and the other mentioned crops impossible without them. Precipitation falls off rapidly from 800 millimeters a year near Cahuasquí to 600 millimeters at La Victoria, 5 kilometers to the east.[51] At a rough estimate, 3 square kilometers may have been irrigated.

TUMBABIRO

Tumbabiro and Las Salinas were alternative names for the same early colonial political unit or pueblo (Figure 6.1). Salt was the major product, but there were also "many cotton fields."[52] These fields cannot be located precisely.[53] The 1639 testament of Felipe de Guzmán, chief of the Salinas pueblo, indicated he had lands near Cahuasquí and maize lands in Urcuquí.[54] In the late 1570s, the area called Yanburo (probably near Hacienda Añaburo, southwest of modern Tumbabiro) was a parcialidad with 159 persons, receiving cotton from Otavalo for processing.[55]

URCUQUI

At Urcuquí (Figures 6.1, 6.4), the Acequia de los Caciques is 14 kilometers long and in 1980–81 provided water to 512 users with a flow of about 160 liters/second.[56] The canal is rustic, with a rustic offtake and passes close

---

48 See his last will and testament, "Testamento de Don Felipe de Gusmán Cazique Principal del Pueblo de Salinas," (1639), Archivo Histórico del Banco Central, Ibarra, Juicios, Paquete No. 10 (1645–1683) ff. 24.

49 Garcés, *Libro de Proveimientos*, 101.

50 See the reference to "una acequia de agua que sale del Río Grande de Cahuasquí para unas huertas que tiene en Puchimbuela," Garcés, *Libro de Proveimientos*, 150.

51 PRONAREG-ORSTOM, *Isoyetas*.

52 Paz Ponce de León, "Relación."

53 Gondard and López believe that air photograph patterns visible below Tumbabiro are suggestive of old, possibly prehispanic, irrigation. See Gondard and López, *Inventario*, site I-156.

54 "Testamento de Don Felipe de Gusmán," AHBC/I.

55 Chantal Caillavet, "Les Chefferies Préhispaniques du Nord de L'Equateur: Formes D'habitat et Organisation Territoriale," *Bulletin Institut Frances Etudes Andines*, 18 (2) (1988), 41–59.

56 "Concession I-73–589: Acequia 'Grande o de Caciques'," (1973–1982), Instituto Ecuatoriano de Recursos Hidráulicos, Ibarra; INERHI (Instituto Nacional de Estudios de Recursos Hidráulicos), *Ríos Palacara, Huarmiyacu, Cariyacu, y Pingunchela, Inventario del Uso de Agua*, Quito, 1980.

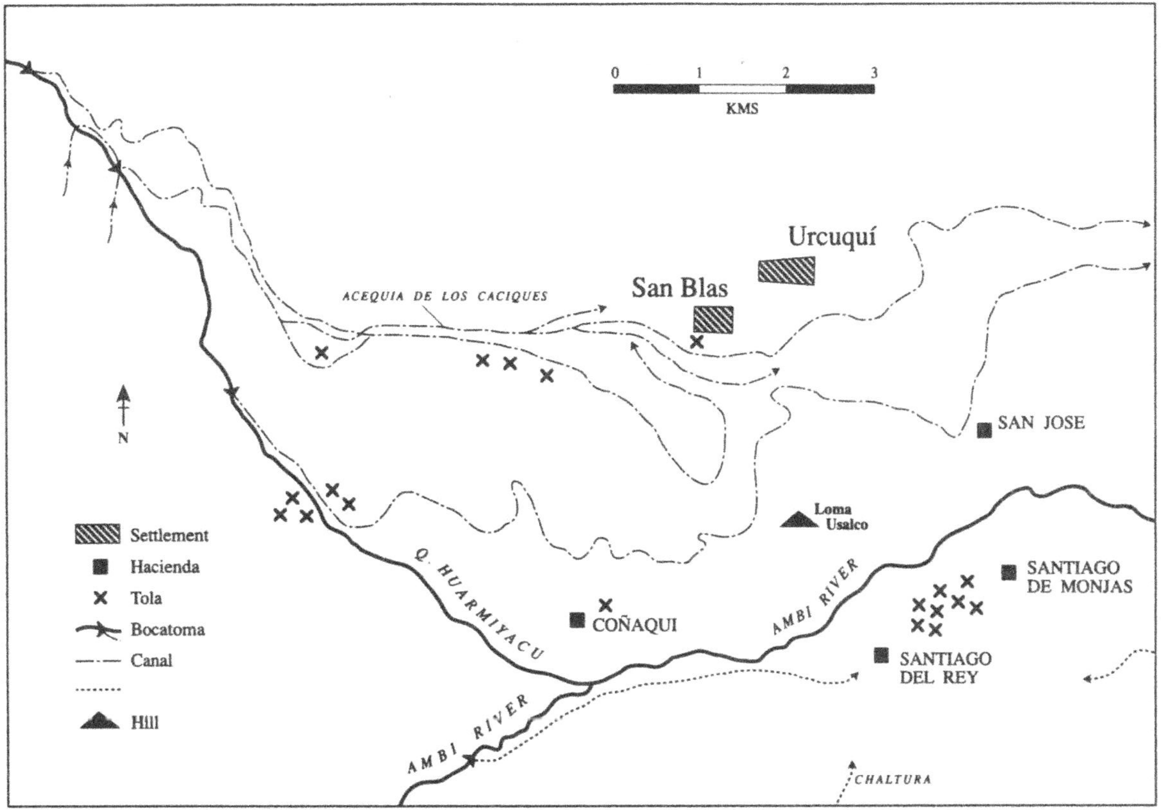

Figure 6.4 Urcuquí Irrigation Zone

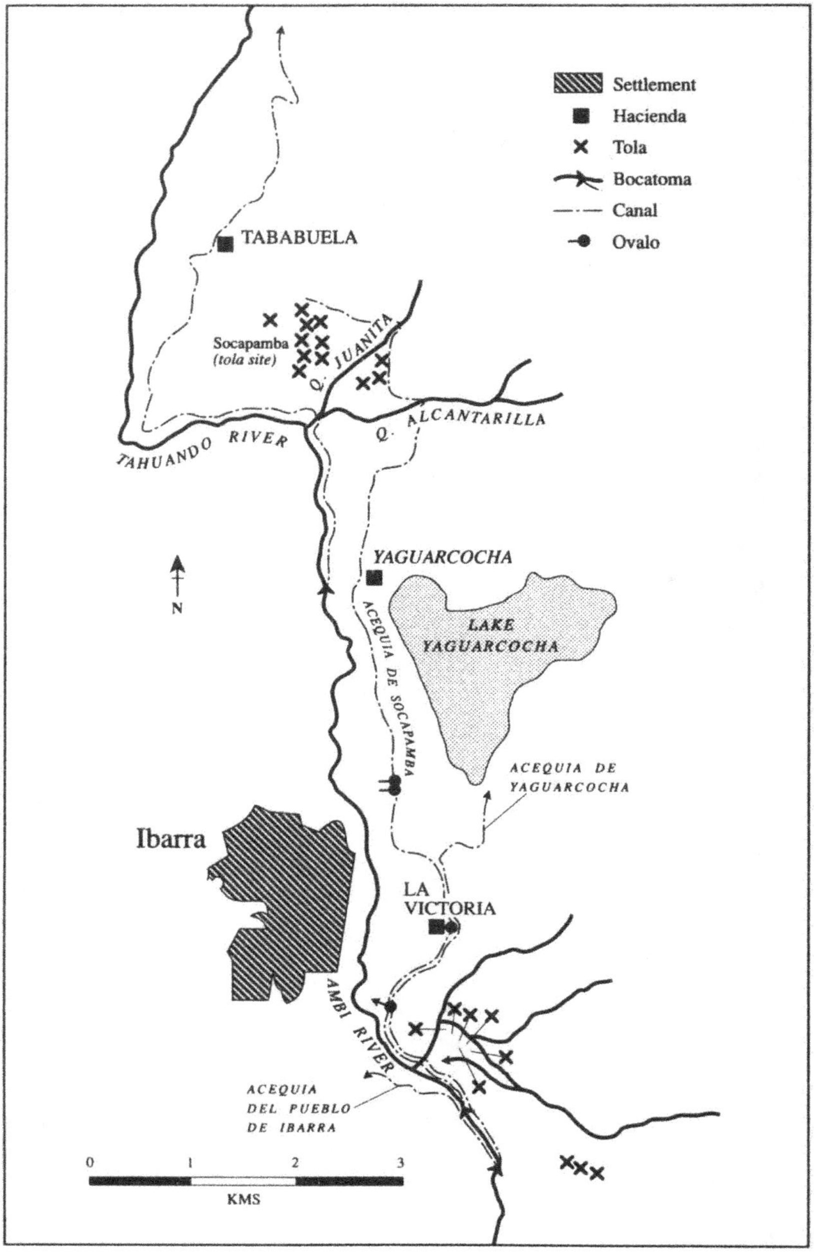

Figure 6.5 Irrigation to the North of Ibarra

by tolas along its route. Elevations vary from about 2,300 meters at San Blas to under 2,100 meters at San José, and the irrigated area is several square kilometers.

Jijón y Caamaño has analyzed colonial documents to suggest that this canal was in use by 1593 for sugar irrigation for the San José hacienda; he also suggests that the abundance of tolas in the area indicate the land must have been used in pre-Hispanic times as well. As late as 1774, the indigenous people of Urcuquí were irrigating their cotton fields with this canal.[57]

Another canal was providing water for the "pueblo" of Cuñaqui (Coñaqui) in 1612.[58] It is interesting to note that both Cuñaquí and Yacelga (Usalco, in the San José area) were parcialidades related to the Otavalo chiefdom in the late 1570s, receiving cotton from the Otavalo chief to turn into tribute cloth; the respective populations were 65 and 413.[59]

After the great 1868 earthquake, the town of Urcuquí attempted to expand its rights over the Acequia de los Caciques. After long conflicts and litigations, in part based on the pre-Hispanic use of canal water, the water rights were transferred away from the San José Hacienda.[60]

While Urcuquí receives about 800 millimeters mean annual precitation, the values fall off with decreasing elevation to perhaps 600 millimeters or 500 millimeters at San José.[61]

TABABUELA

To the north of Ibarra is another subtropical dry area with tolas and old canals (Figures 6.1, 6.5). The Socapamba canal today (1985) derives water from the Yaguarcocha canal. Originally it derived water from the Tahuando River and dropped along the Juanita ravine to irrigate the Tababuela sugar cane fields. In 1977 the canal irrigated 50 hectares at La Victoria and 250 hectares at Tababuela.[62]

Above the old offtake there are two tola sites.[63] In Socapamba, where the canal drops to lower elevations, there are 60 tolas, including 2 ramp

57 Jacinto Jijón y Caamaño, *Historia de la Llamada "Acequia de los Caciques" en Urcuquí*, Quito: Editorial Unión, No Date.

58 [...Don Maseo Tulcamaca Cacique Principal del Pueblo de Tulcan...], (1612), Archivo Histórico del Banco Central, Ibarra, Notarias, Libro 2º , 60 r.

59 Caillavet, "Chefferies Préhispaniques," 51.

60 Anonymous, *Cacique Urcuquí vs. Gamonalismo*, Quito: Editorial Quito, 1946.

61 PRONAREG-ORSTOM, *Isoyetas*.

62 "Inventario de los Usos de Agua, Subcuenca del Rio Ambi," (1977), Instituto Nacional de Estudios de Recursos Hidráulicos División de Recursos Hidrológicos, Quito.

63 The northern site, in the La Matilde hacienda, has 8 hemispheric tolas and a larger tola to the west. Farther south, in the Hacienda Lulunquí Madera, there are three

tolas. The ceramics here represent two traditions, one predating tola construction, the other corresponding to the last phase of tola construction, 1250–1550 A.D.[64]

The lands of Tababuela do not show evidence of tolas today, but much pre-Incaic ceramics have been found.

A 1592 litigation over a piece of land by Don Marcos Velasqués, "principal Indian and mayor ... of the town of Caranque," suggests that this area was irrigated and cultivated by Indians of Caranqui:

> ...a piece of land ... which borders on lands of ... the aforesaid Don Marcos ... and on the upward side on the canal which goes from the mill of Ynez Rodríguez to the olives of San Augustín and on the lower side the great river of Carangue and with the tola which is by the aforesaid river...[65]
>
> The "olives of San Agustín" correspond with San Pedro de los Olivos, today El Olivo; the great river is the Tahuando.[66]

Much of the area irrigated by the Socapamba canal was originally acquired by the Jesuits, but the Augustinians gradually acquired them and by 1619 the Augustinian haciendas of Yaguarcocha and Tababuela were complete.[67] There is a reference of 1737 to a canal which derived water from the Tahuando River and went to the Natabuela (=Tababuela) hacienda of the Augustinians.[68]

Another document of 1803 refers to a canal "of much cost:"

> for the benefit of this hacienda ... waters were derived from the Tahuando River and ... went to irrigate the lands of Tababuela.[69]

It appears the earthquake of 1868 did not substantially change this canal, since in 1874 there is a reference to "the canal which conducts the water for the aforesaid Hacienda La Victoria and for Tababuela."[70]

---

tolas. See José Echeverría, *Proyecto: Sistemas de Irrigación Antiguos en la Sierra Norte del Ecuador*, Otavalo: Instituto Otavaleño de Antropología (Manuscript in possession of the author), 1986, for a discussion of the ceramics and lithics.

64 John Stephen Athens, *El Proceso Evolutivo en las Sociedades Complejas y La Ocupación del Período Tardío - Cara en los Andes Septentrionales del Ecuador*, Otavalo: Instituto Otavaleño de Antropología, 1980.

65 [...Yo don marcos velasques indio principal y alcalde ordinario de pueblo de carangue...], (1592), Archivo Histórico del Banco Central, Ibarra, Notarios, Libro 1º, ff. 18.

66 Cristobal Tobar Subía, *Monografía de Ibarra*, Ibarra: Prensa Católica, 1950.

67 "Libro Mayor Títulos y Fondos de este Convento de Nuestro Señor San Agustín, Volumen VIII," (No Date), Archivo Histórico de San Agustín, Quito.

68 Juan de Dios Nabas, *Ibarra y Sus Provincias de 1534 a 1932*, Quito: Imprenta del Clero (2 Volumes), 1934, Volume 2, 22.

69 [Concerning the Water Supply of the Acequia de Tababuela], (1803), Archivo del Banco Central, Ibarra, Juicios, Paquete 222.

70 [Document of Sale of Hacienda La Victoria to Sr. Miguel Teodoro Muñoz], (1874), Archivo Histórico del Banco Central, Ibarra, Notario 2º, ff. 148,149.

Thus a canal existed to irrigate the Tababuela lands since the sixteenth century, and the lands themselves exhibit the tolas and ceramic scatter to indicate pre-Hispanic use. The low precipitation at Tababuela (600 millimeters or so) also suggests the probability of irrigation to enable settlement.[71] The lands are located between 1,700 and 1,900 meters elevation, with the tola site of Socapamba just below 2,200 meters.

AMBUQUI

The Ambuquí River flows through a gently sloping valley from 2,000 to 1,700 meters elevation; the river has water all year, although the summer quantity is low. A series of springs contributes to the river's flow.

The mean annual precipitation is very low (about 500 millimeters) so irrigation is necessary for worthwhile production of most crops. Today the valley is controlled by the two haciendas Irumina and Iruminita, and other lands by the communities of San Clemente and Ambuquí. Six traditional rustic canals are in use, and irrigate approximately 240 hectares. (Figures 6.1, 6.6).

In this valley tree cotton is still grown, suggestive of the aboriginal use of these side valleys. Other crops include sugar cane, obo (*Spondius purpurea*), grapes, avocados, plantains, and beans.

The valley exhibits a number of tolas plus the "Tablón" archaeological site with pre-Incaic ceramics. The upper valley has the toponym Guaranguí, apparently the same as the pueblo of "Guarangue" which existed prior to the 1572 resettlement of regional groups in Pimampiro.[72] The ethnic lord of this or another settlement in this valley was named Ambuquí; he had forty tributary Indians under his control. By 1582, the valley was being farmed by Indians resident in Pimampiro for coca and cotton; there were also vineyards.[73]

A drought-induced litigation over water in 1661 provides more information on the irrigation history of this valley.[74] A witness deposed that:

> ...since antiquity [the Indians had] derived ... canals and after the Spaniards had entered that valley they had taken away the aforesaid water from the natives...

One witness mentioned that the valley had six bocatomas, corresponding to the present situation (1985), an extraordinary example of irrigation stability. Indians, many of them born in the valley but identifying themselves as pertaining to San Antonio de Ibarra and Caranqui, were irrigating coca, cotton, chili peppers, maize, fruit trees, and anise, but were concerned with

71 PRONAREG-ORSTOM, *Isoyetas*.

72 Caillavet, "Chefferies Préhispaniques," 45.

73 Borja, "Relación," 250.

74 "Repartición de Agua de Ambuqui y Repartimiento de las Aguas," (1661), Archivo Histórico del Banco Central, Ibarra, Juicios, Paquete 6 ff. 12.

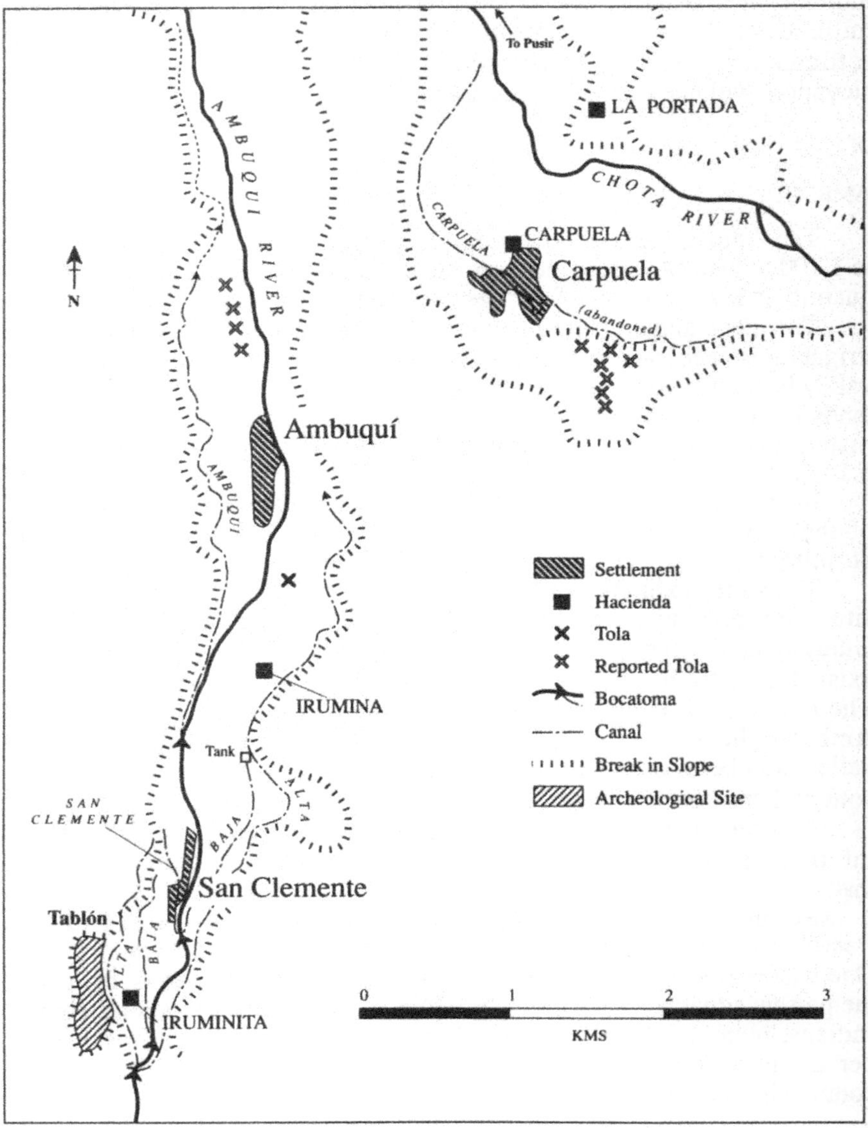

Figure 6.6 Irrigation at Ambuquí

water robberies by a black on behalf of his Spanish owner. A Spanish witness deposed that for three generations his family had irrigated his farm, and that the farm in turn had been acquired from Augustinian fathers; but no Spaniards came forward to claim the canals were of Spanish origin.

COANGUE

To the north of Pimampiro, the Mira or Chota River has cut a deep canyon; the flood plain and the adjacent fluvial terraces and fluvial/colluvial fans are easily irrigated from the river and its tributaries (Figures 6.1, 6.7, part of Figure 6.6, and Figure 6.8). There is no meteorological station, but extrapolations from nearby stations suggest that 300–400 millimeters mean annual rainfall may be received.[75] The canyon floor has an elevation of 1,600 meters, apt for a wide range of tropical crops; today sugar cane, plantains, potatoes, avocados, cucumbers, obos, grapes, sweet potatoes, tomatoes, and manioc are grown. Maize is rare. Today the region is irrigated by a system of new canals constructed in the 1970s by INERHI at a cost of more than a million dollars; about 1,600 hectares are irrigated.

The area is inhabited by the descendents of African slaves brought to farm the colonial haciendas which came to dominate the valley. Originally the area was inhabited by diverse Indian groups. We have a good description of this valley in 1582, when the valley was called Coangue:[76]

> ...the aforesaid valley [Coangue] is formed by a great river called the Mira ... There used to be four pueblos settled in this aforesaid valley of Coangue on the banks of the river, and the inspector general [visitador general] ... grouped and settled them in this settlement of Pimampiro...
>
> These Indians have ... their fields of coca and cotton and maize and other crops in this aforesaid valley of Coangue ... It is a very fertile valley ... although unhealthy some times of the year ... They irrigate the fields of coca and cotton which are in the valley with the water of the big river...

By 1582, there were 888 souls still alive in Pimampiro of those who had been resettled from the valley. In addition, there were more than 1580 tributary Indians who were "foreigners" who had come to cultivate coca or oversee their workers. These people came especially from the north (Pastos) but also from Otavalo, Caranqui, Latacunga, and Sigchos.[77]

The older canals of the Coangue valley include the Acequia del Río Sur (abandoned), Acequia Madre de la Playa, Acequia del Río Norte (abandoned), Apaquí, Changona (2 canals), Chalguayacu (3 canals), Carpuela (abandoned), El Refugio, Pusir (abandoned), Pusir Chiquito (abandoned) and Tumbatu (abandoned). The canals may have originally irrigated 930 hectares.

---

75 PRONAREG-ORSTOM, *Isoyetas*.

76 Borja, "Relación," 248–249.

77 Borja, "Relación," 252.

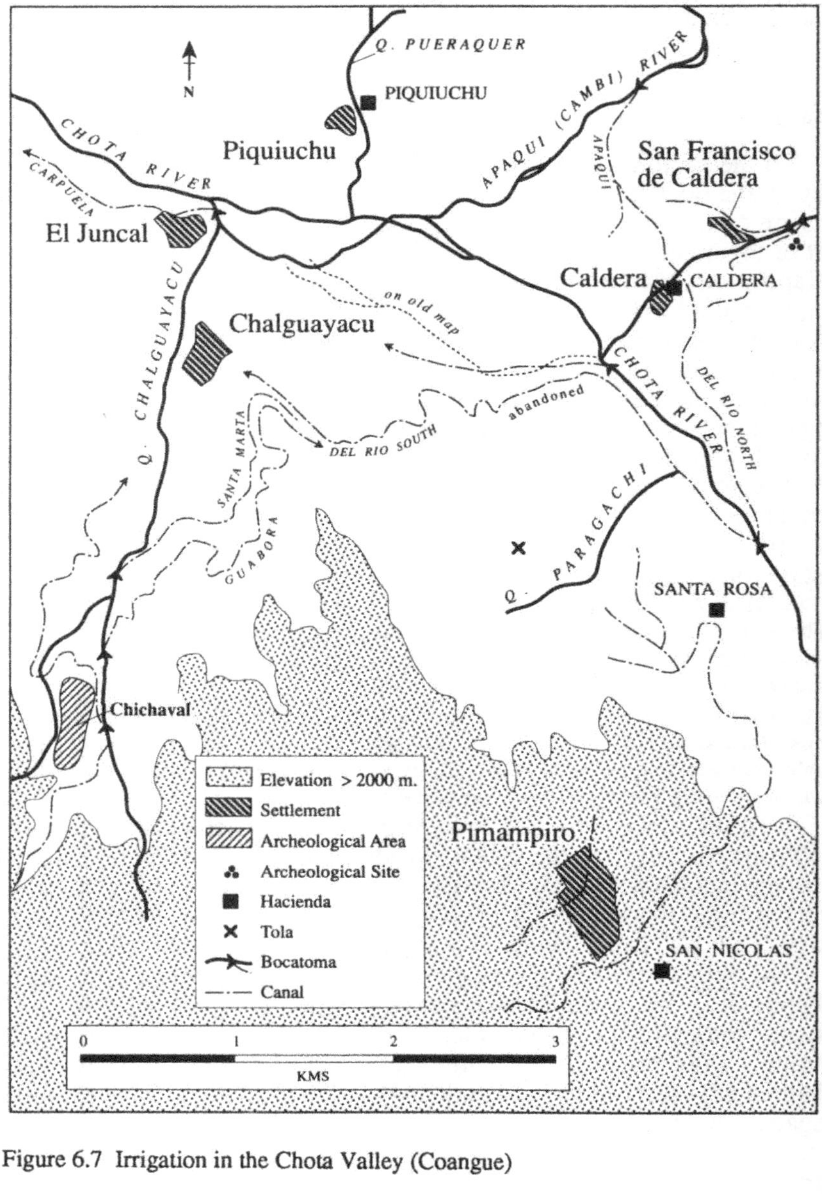

Figure 6.7 Irrigation in the Chota Valley (Coangue)

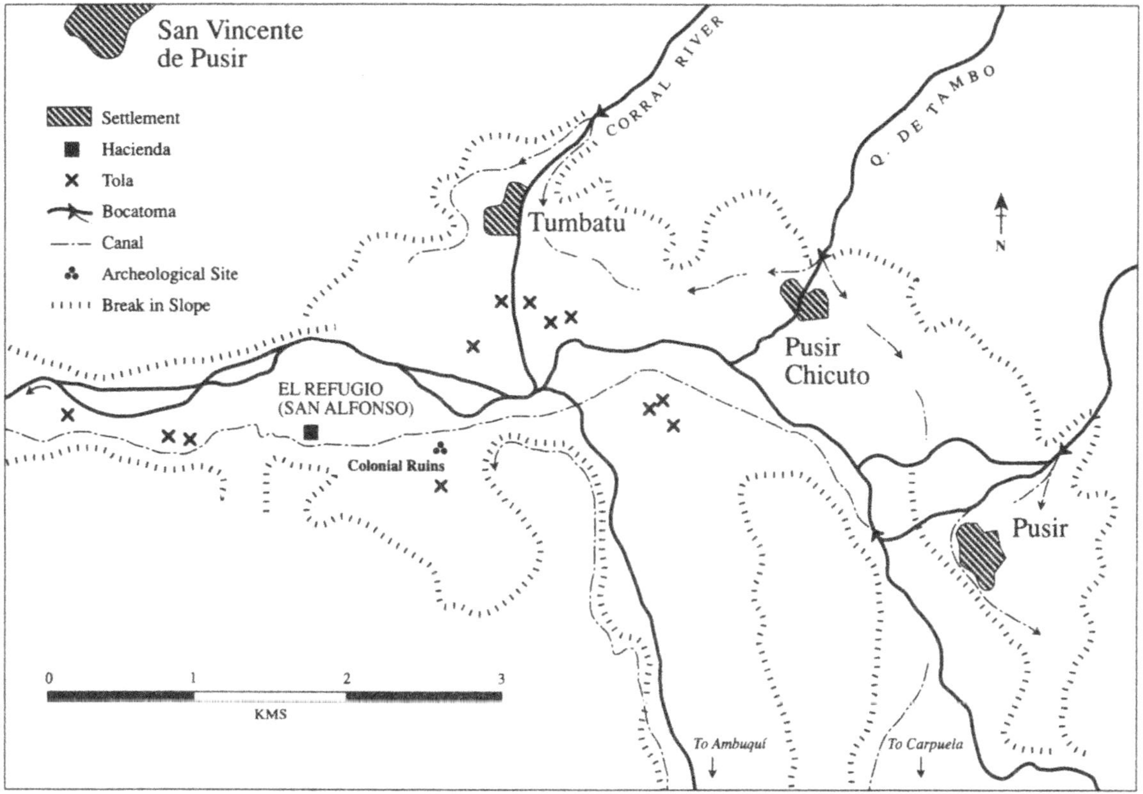

Figure 6.8 Irrigation of the Lower Chota Valley

Early testaments analyzed by Chantal Caillavet[78] allow us to learn how the coca was grown. It was raised in lines (*rayas*) often interspersed with fruit trees such as avocados. The *raya* was used as a term of measurement for legal purposes; thus it probably was not an impermanent huacho but a fixed landscape feature, perhaps related to irrigation. One document distinguishes between coca fields and camellones, however, so the *raya* should probably not be confused with raised fields (see below). Unfortunately no features have survived which can be identified with the *raya*; coca cultivation was outlawed in Ecuador in late colonial times.

La Concepcion

A number of irrigation canals irrigate the gentle slopes at La Concepción (Figure 6.1) between 1,350 and 1,500 meters elevation. Mean annual precipitation is about 600 millimeters,[79] and the major irrigated area is about 3 square kilometers in area.

This area exhibits tolas on both sides of the river. No known llajta corresponds with this site, which may have been one of the communities resettled at Pimampiro. The low precipitation suggests an irrigated cotton or coca center.

The Guayllabamba Basin

Pre-Hispanic irrigation in the subtropical lower Guayllabamba basin (Figures 6.1, 6.2) is a bit more difficult to define than farther north, as this region seems to have been somewhat more affected by the Inca and Spanish conquests. Guayllabamba, for example, was a major river crossing and fortress center at the Guayllabamba River, not located at the present site of the city of Guayllabamba but at Pueblo Viejo to the west. It probably had little pre-Hispanic agricultural importance, but in the sixteenth century there were some irrigated orchards in the subtropical lowlands next to the Pisque and Guayllabamba Rivers.[80]

A bit further downstream, the Otavalo chief had coca fields at Puéllaro (Figure 6.1) in 1609.[81] Tolas are present in the nearby irrigated zone of Alchipichi (La Libertad), at an elevation extending from 2,100 down to 1,800 meters. Close to this site, there is a bridge (pasarela) for an old Quito-Otavalo route which also passes by the fortress of Rumicucho.

---

78 Chantal Caillavet, "Toponimia," 15–16.

79 Mean annual precipitation at Estación Carchi, 1965–1973, was 579 millimeters. See PRONAREG-ORSTOM, *Estudio Hidro-Meteorológico*, 54.

80 See Paz Ponce de León, "Relación," 237–238.

81 Caillavet, "Caciques de Otavalo," 47.

Even farther downstream, there is a reference to canal irrigation at Ambuela in 1583:[82]

> ...three caballerías of land ... in a valley which is called Tutambuela above the River of Guayllabamba, in the highway which goes to Old Perucho and the settlement which has been made called Cofo, as one comes from Perucho to [Quito] ... in a deep depression up to the cattle ranch [*hato*] which goes to the town of Pillaharo [Puéllaro], from *the high canal* on down ... (italics added)

Today, Ambuela is on the road going to Atahualpa and the Hacienda Mojanda; the old Perucho could have been in this area.[83]

Conclusion: Subtropical Irrigation

The various sites documented above suggest that cotton was grown below 2,300 meters and coca below 2,000 meters, under precipitation regimes of 600 millimeters per year down to 300 millimeters per year. Maize and fruit trees were frequently also grown in these specialty crop sites. Many of the irrigated areas appear to be equivalent to about 2.5 to 3.0 square kilometers, and the political units controlling the irrigated lands in every case were associated with larger paramount chiefs at higher elevations.

Costs and Benefits of Pre-Hispanic Irrigation

The benefits of pre-Hispanic irrigation in terms of soil fertility were presumably comparable to those realized today. The costs were different primarily in terms of canal maintenance. If 2.7 times the present labor input were required, a district such as Pimampiro with a 20 kilometer long canal and 600 hectares irrigated would require about 7 person-days per hectare per year for each canal cleaning operation.[84] This is not a particularly large input by pre-Hispanic standards, and it might be expected that irrigation in pre-Hispanic times would be at least as attractive in terms of the alternatives as in modern times (see Chapter 8).

---

82 Garcés, *Libro de Proveimientos*, 58.

83 Hacienda Mojanda was the site of the "pueblo" called Axangui in the sixteenth century; some of its citizens were subject to the authority of the chief in Otavalo, and it is conceivable that the area functioned as an irrigated cotton supply base. See Garcés, *Libro de Proveimientos*, 115, 89, 13; and Caillavet, "Chefferies Préhispaniques."

84 200 x 7.5 x 2.7 = 4,050 person-days labor for canal cleaning; this divided by 600 hectares is 6.75 person-days per hectare per year.

Chapter 7

PREHISTORIC RAISED FIELDS

"Raised fields" are planting beds (or mounds or ridges such as wachukuna) (Plate 7.1). Unlike drained fields, the planting surfaces of raised fields have been substantially raised by continued deposition of spoil from the adjacent ditches. In this chapter, however, the term "raised fields" will be used only when the distance between adjacent raised field midpoints (wavelength) is from 3 to 10 meters, and the distance from field top to ditch bottom (amplitude) is 1 meter or more;[1] wachukuna are not included. This distinction is based on functional considerations; raised fields (in the sense used here) always have standing water in their intervening ditches, and consequently have special soil fertility, subirrigation and frost management functions which wachukuna in Ecuador do not possess.[2]

Discovery of Raised Fields

Sixteenth century Spaniards appear to have observed abandoned raised fields in the areas of Bogotá, Colombia,[3] and perhaps near Otavalo;[4] there

---

1 The terms wavelength and amplitude were introduced to raised field studies by Martin Parry, "A Typology of Cultivation Ridges in Southern Scotland," *Tools and Tillage*, 3 (1976), 8.

2 This usage sidesteps the issue of developing a taxonomy of raised field forms. See Kent Mathewson, "Taxonomy of Raised and Drained Fields: A Morphogenetic Approach," in Ian Farrington (ed.), *Prehistoric Intensive Agriculture in the Tropics*, Oxford: BAR International Series 232, 1985, 835–845. Wachukuna with water in the intervening ditches do exist in southern Peru; see footnore 37, below, and Karl Zimmerer, "Rural Society and Wetland Ecology in the Agricultural Transformation of a Highland Peruvian Region," *Annals of the Association of American Geographers*, (In Press).

3 See the reference to "terrenos ... surcados por anchos camellones que son vestigios de antiguos cultivos" by Joaquín Acosta, *Compendio Histórico del Descubrimiento y Colonización de la Nueva Granada en el Siglo Décimo Sexto*, Paris: Imprenta de

is also mention of raised fields in the native wills of the Otavalo area, where they are referred to as *camellones* in Spanish, and as *pigal*, *bigal*, or *nigal* in the local pre-Inca language.[5]

Despite these mentions, Andean raised fields were forgotten until rediscovered in the twentieth century by Uhle, Smith, Denevan, Eidt, and Broadbent.[6] In the last two decades, a number of sites have been reported, and increased emphasis has been placed on the interpretation of the functioning of these features.

In Ecuador, the modern rediscovery of raised fields is due to Roy Ryder, who in 1969 found the abandoned fields in the plains of Paquiestancia and Cayambe (Figure 7.1).[7] In 1970, more abandoned raised fields were found in the San Pablo flats by Roy Ryder and Alfred Siemens (Figure 7.1).[8] In 1980, raised fields were discovered by the author in the Quito flats (Figure 7.1),[9] and (independently) by the author and Pierre Gondard and Freddy López in the area of Otavalo-Pinsaquí. Gondard and López have also found abandoned raised fields at San José de Minas and a number of small flats to the west of Otavalo (Figure 7.1).[10]

Beau, 1848, 204; Sylvia Broadbent, "A Prehistoric Field System in Chibcha Territory, Colombia," *Ñawpa Pacha*, 6 (1968), 140.

4 See the 1573 reference to abandoned fields ("labores de sementeras") near Otavalo, Paz Ponce de León, "Relación," 235.

5 See the passages in last wills and testaments cited by Caillavet, "Toponimia," 12.

6 In 1923, Max Uhle briefly mentioned "ancient cultivated fields of a special type, in the form of slightly elevated beds more than a meter in width" for the Bolivian shore of Lake Titicaca and the southern highlands of Ecuador; see Max Uhle, "The Aims and Results of Archaeology [1923]," in John Rowe (ed.), *Max Uhle, 1856–1944*, Berkeley: University of California Publications in Archaeology and Ethnography, 1954, 86. In 1950 and 1951, Robert Eidt studied modern raised fields in the high wet plain around Bogotá, Colombia; the plain also has abandoned raised fields. See Broadbent, "Prehistoric Field System;" Robert Eidt, "Aboriginal Chibcha Settlements in Colombia," *Annals of the Association of American Geographers*, 49 (1959), 374–392; Robert Eidt, "Detection and Examination of Anthrosols by Phosphate Analysis," *Science*, 197 (1977), 1327–1333; and Eidt, *Advances*. For an early discovery of raised fields on the Lake Titicaca plain, see Clifford T. Smith, William M. Denevan and Patrick Hamilton, "Ancient Ridged Fields in the Region of Lake Titicaca," *The Geographical Journal*, 134 (1968), 353–367.

7 Ryder, "El Valor."

8 The first publication of this site appears to be in Athens, *El Proceso Evolutivo*.

9 Reported in Gregory Knapp, "El Nicho Ecológico Llanura Húmeda, en la Economía Prehistórica de los Andes de Altura," *Sarance*, 9 (1981), 83–96. A more detailed report of the site is found in Gregory Knapp and Roy Ryder, "Aspects of the Origin, Morphology, and Function of Ridged Fields in the Quito Altiplano, Ecuador," in Janet P. Darch (ed.), *Drained Field Agriculture in Central and South America*, Oxford: B.A.R. International Series 189, 1983, 201–220.

10 Knapp, "Nicho;" Gondard and López, *Inventario Arqueológico*.

Plate 7.1 Raised Fields at Cayambe (photo by author)

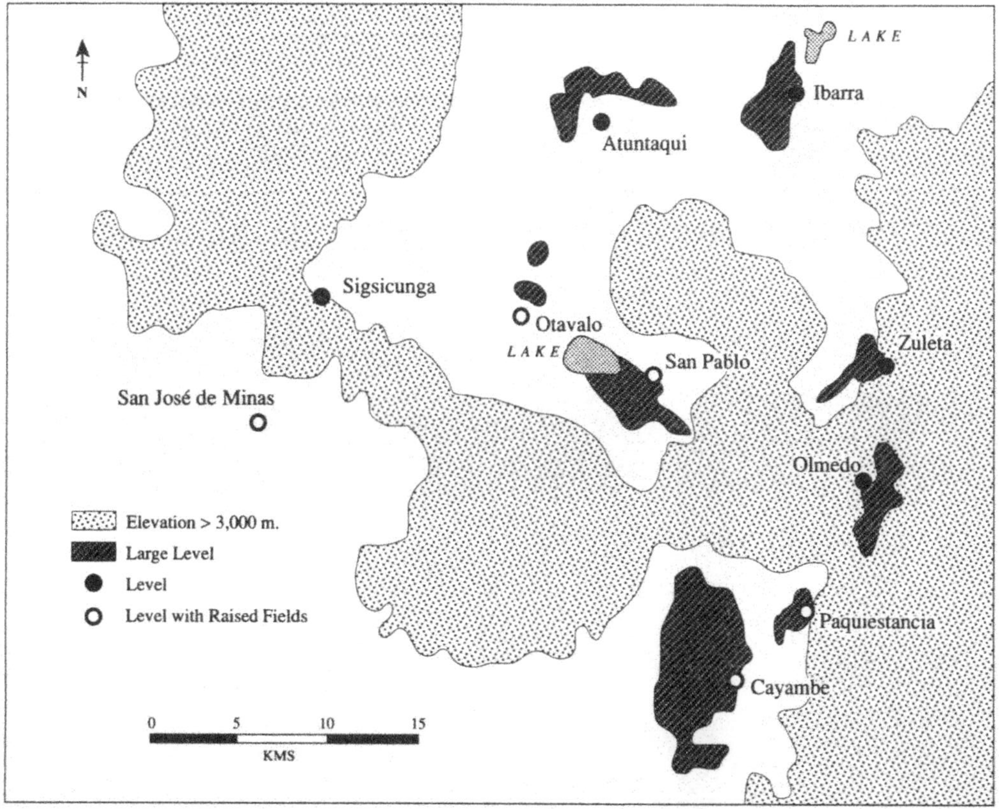

Figure 7.1 Location of Large Flats and Raised Field Sites

The Quito Site

Raised fields have been found northeast of Chillogallo, in the grassy plain south of Quito (Figure 7.1).[11] About 8 hectares of pasture with raised fields may be observed on available (1968) air photographs; these features undoubtedly occupied a greater area than this but they have been largely covered by ash falls, or destroyed by urban sprawl and pasture improvement. The total area of the plain south of Quito is 18 square kilometers, and since true raised fields are not associated with slopes, this can be considered the maximum possible extent of the original complex. The elevation of these flats is about 2,850 meters.

The wavelength, or distance from ridge center to ridge center, is fairly consistently 3.5 meters. Excavations have indicated that ditch width was originally equal to ridge width. The original distance from ditch bottom to ridge top was over 0.8 meters. Ridge length varied greatly but was typically 25 to 75 meters. If only the ridge tops were cultivated then the width of the ditches and ridges implies that the original cultivated surface was half of the total area of the plain.

No ceramic or other cultural material was recovered from our excavations of the ridged fields. However, the site is located close to the late Prehistoric Chillogallo settlement excavated by Echeverría.[12] The latter was a longitudinal settlement of family nuclei located along the edge of the flats. Abundant burials are evidence of a dense population and cultural remains indicate an affinity with more northerly cultures such as Cayambe, also known to be associated with raised fields.

Post-conquest accounts describe the flats to the south of Quito as being marshy and brushy, serving as pasture for cattle. The indigenous population was described as living on the slopes, farming maize and potatoes.[13] The scientist-traveller Francisco José de Caldas described the same flats as marshy and flooded in the early 19th century.[14]

A resident of Barrio Santa Rita who was interviewed during our excavations declared that the area of surviving raised fields was marshy until drained 30 or 40 years ago. Local farmers deny that the Santa Rita raised fields have been farmed in memory.

---

11 Knapp and Ryder, "Aspects."

12 See José H. Echeverría, *Contribución al Conocimiento Arqueológico de la Provincia de Pichincha: Sitios Chilibulo y Chillogallo*, Quito: Facultad de Ciencias de la Educación, Pontífica Universidad Católica del Ecuador (Thesis), 1976.

13 Anonymous, "Cibdad," 212, 223, 227; Pedro Rodríguez de Aguayo, "Descripción de la Ciudad de Quito y vecindad de ella por el arcediano de su iglesia, licensiado Pedro Rodríguez de Aguayo [No Date]," in Marcos Jiménez de la Espada (ed.), *Relaciones Geográficas de Indias - Perú, Volume 2*, Madrid: Biblioteca de Autores Españoles 184, Ediciones Atlas, 1965, 202, 204.

14 Francisco José de Caldas, *Bosquejos Inéditos*, Bogotá: Seminario del Nuevo Reino de Granada, Volume 3, Re-edited by Biblioteca Popular de Cultura Colombiana, Historia 9, Editorial Kelly, 1942, 36.

Mean annual precipitation of this site is 1,222 millimeters (Quito Observatory); current land use was pasture in process of conversion to suburban house sites; deep drainage ditches maintained the water table below 1.5 meters.

Preliminary probes of raised field ditches and tops with a soil auger indicated that the ditches contained about half a meter of fill, composed mainly of a grey volcanic ash not found in the ridges. A trench was excavated perpendicularly across one of the ditches revealing in greater detail the nature of this fill. It was discovered that the ditch had received volcanic ash from at least two separate eruptions, the first coinciding with the date of field abandonment.

Figure 7.2 helps illustrate a proposed reconstruction of the filling process. Before construction of the raised fields, the flats had a brown-black horizon, A, and a bright black horizon, B. Raised field construction and agricultural operations destroyed the original A horizon which was mixed with B material to produce a slightly lighter black loam on the raised field tops, C. Most of the ditches probably contained standing or flowing water for subirrigation, frost amelioration, and accumulation of silt and muck for fertilization.

A layer of white volcanic ash, V was deposited on the landscape and was blown and washed from the ridges into the ditches. At this time, the raised fields were abandoned. They were not abandoned earlier since there is no fill beneath the ash layer. Now that the ditches were without maintenance, material quickly silted in from the adjacent raised field sides, D. This material has a higher organic matter content than the adjacent ash materials. In addition, a brown-black A horizon began to develop where the black C and D materials were more stable.

Additional layers of volcanic ash fell on the plain and were transported into ditches. With time, these ash layers were capped by material transferred from the ridges to the ditches by erosion and mass movement, E. This layer exhibits the highest phosphate concentration of any of the ditch fill layers; at 78 kilograms per hectare phosphate, it is also much richer than the adjacent raised field. Since phosphorus is the single most important limiting factor for agriculture in these soils, most of the material accumulating in the ditches between raised fields would have been valuable for promoting soil fertility on the planting surfaces. The only exception would have been the relatively phosphorus-poor recent ash falls.

With more time, the lighter colored A horizon which was developing over the entire landscape became more pronounced; dark organic colorants were leached into the underlying horizon. The fluctuating water table promoted the formation of iron/manganese concretions in the developing V horizon. The boundary between layer E and the adjacent raised fields became obscured by the formation of the new A horizon. As the D horizon was formed from slumped raised field material, the ditch boundary is also observed with difficulty at the D/C contact. Only the edges of the first ash deposit clearly indicate the original sides of the ditch.

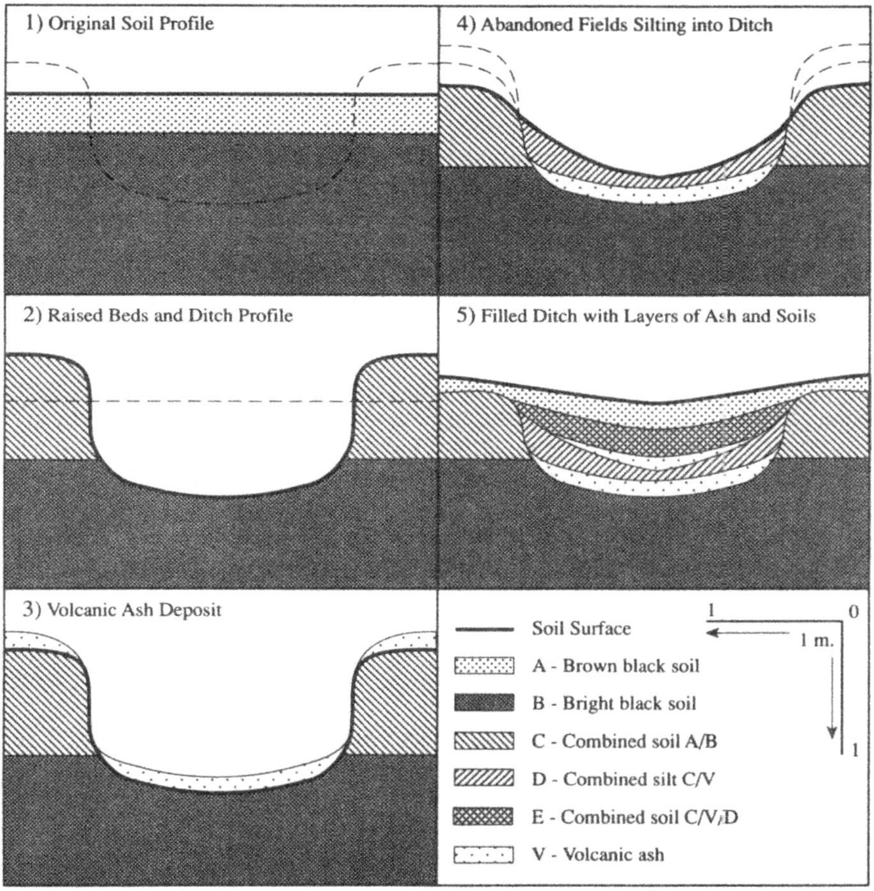

Figure 7.2 Raised Field Ditch Filling Sequence, Chillogallo

San Pablo Site

Raised fields are found in the San Pablo flats to the southeast of Otavalo.[15] Air photographs indicate many areas of raised fields and there are many in a good state of preservation. Using air photographs and interpolating, at least 512 hectares are in raised fields,[16] while the total area of flats (and potentially the total area in raised fields) was 1,170 hectares. The local elevation is 2,700 meters.

The Itambi River has cut across some areas of raised fields, enabling the determination of the original cross section. In amplitude the San Pablo fields are roughly comparable to those at Chillogallo; wavelength is somewhat longer, about 3 to 5 meters. Orientation is diverse, with a checkerboard pattern predominating; orientation and arrangement would have been counterproductive if drainage were the primary reason for raised field construction. On the other hand, the arrangement would have allowed maintenance of standing water in the ditches if appropriate temporary check dams had been constructed.

The San Pablo flats, according to one of the current landowners, became latifundia property during the colonial period. Through changing ownership this land came to be part of Hacienda Cusín. In recent years, the hacienda has come to be further subdivided between a cooperative, small proprietors, a hotel, and one larger farm (Hacienda La Vega). In 1981 I interviewed the owner of Hacienda la Vega, older employees of Hacienda Cusín, other landowners in the flats, and subsistence farmers on the neighboring slopes; all alleged that the raised fields were prehistoric, and that previously almost the entire plain was full of them. Hillslope subsistence farmers called the features "Inga wachukuna" (Inca wachukuna) and remarked that the Spanish had taken the good flat land from their ancestors.

In 1981 raised fields in the San Pablo plain were excavated.[17] In one location, raised fields were structurally integrated into the sides and ramp of a small ramp tola; excavation of the ramp failed to disclose any buried surfaces such as would have been produced if the raised fields were constructed after the ramp tola. Thus it appears that the ramp tola and the surrounding raised fields were all constructed about the same time. Broken pieces of late pre-Inca Carchi Negative or Capulí ware were found in the

---

15 Gregory Knapp and William Denevan, "The Use of Wetlands in the Prehistoric Economy of the Northern Ecuadorian Highlands," in Ian S. Farrington (ed.), *Prehistoric Intensive Agriculture in the Tropics*, Oxford: BAR International Series, 1985, 185–207.

16 Gondard and López, *Inventario Arqueológico*, 148; this value is higher than previous estimates by Ryder and Athens, in part due to Gondard and López's methodology of inferring raised fields for areas between features visible in photographs.

17 María del Carmen Molestina, "Investigaciones Arqueológicas en la Zona Negativo del Carchi o Capulí," *Cultura (Quito)*, 7 (21a) (1985), 31–82.

raised fields as well as in the associated ramp tola; a carbon sample from this tola was insufficient for a proper radiocarbon date, but elsewhere similar ramp tolas have yielded dates after 1250 A.D.[18] Several pieces of carbon were recovered from the fill in a raised field ditch next to the ramp of this tola. The composite sample yielded a radiocarbon date of 500 ± 100 years B.P.[19] Since the ditch fill postdates abandonment of raised field mucking practices, this date is consistent with abandonment of mucking during the Inca or Spanish conquests.

It is also pertinent to observe the presence of the toponym Pijal in the San Pablo area, which in the language of Otavalo referred to a field ridge (*camellón*). There are early colonial references to camellones in the testaments of Indians, but it is unclear whether by this time the ridges were still being mucked; it is also unclear whether the "camellones de papas" mentioned in wills refer to raised fields or simply to potato wachukuna.[20]

Cayambe Site

The largest of the northern Ecuadorian raised field sites is that of Cayambe. Here is a large flat of about 4,230 hectares at 2,800 meters above sea level. Air photo interpretation indicates up to 1,237 hectares were reclaimed by raised fields,[21] and indeed the entire area may have at one time been so used.

The Cayambe site presents examples of raised fields with an amplitude of about 1 meter, wavelengths of 3.5 to 7 meters, and with variable lengths, orientations, and patterns. Cayambe is unique in having numerous semi-circular embankments en echelon apparently designed to impound water.[22] These are reminiscent of similar embankments from Lake Titicaca[23] and from Chilca on the Peruvian Coast.[24]

The Cayambe raised fields are associated with a pre-Hispanic ramp tola site at the eastern edge of the plain,[25] as well as with another site to the east of the present city of Cayambe which reportedly once also had a ramp tola[26]

---

18 Athens, *El Proceso Evolutivo*, 135–137.

19 Sample QL-1747, Quaternary Research Center, University of Washington, Seattle.

20 Caillavet, "Toponimia," 12.

21 Gondard and López, *Inventario Arqueológico*, 148.

22 Batchelor, "Camellones."

23 Smith, Denevan and Hamilton, "Ancient Ridged Fields."

24 Knapp, "Prehistoric Flood Management."

25 Athens, *El Proceso Evolutivo*, 266–267.

26 Jacinto Jijón y Caamaño, *Contribución al Conocimiento de los Aborígenes de la Provincia de Imbabura en la República del Ecuador*, Madrid: Blass y Cía., Estudios de la Prehistoria Americana 2, 1914, 293–294.

with late pre-Inca ceramics.[27] Prehistoric burials have been found in the ditches between raised fields on the Cayambe plain.[28]

Ibarra Site

To the south and west of Ibarra is a large area of flat land with soils similar to those of the raised field sites mentioned above. Although no raised fields are visible today, there is some evidence that the region once had them.

In the Ibarra flats, a 1635 litigationrefers to

> ...these Indians ... in possession of some lands ... inherited from their ancestors ... called Pigalquí which are swampy and marshy...[29]

The toponym Pigalquí contains the element "pigal" which alludes to ridges or raised fields in the Otavalo language. There is also a 1595 document which refers to "diez camellones" in the area of Caranqui to the south of Ibarra.[30]

The Ibarra flats were probably severely effected by the Inca conquest and the Inca slaughter of the inhabitants of this area.[31] By 1573, there is a reference to "vacant and abandoned lands" here where a settlement of Spaniards could be implanted.[32] The town of Ibarra was established about 1606 as such a settlement, and much of the flats were included in the common pasture (*ejido*) of the new city.[33]

Other Sites

Well preserved raised fields occur in association with ramp tolas on the Paquiestancia flats at 2,940 meters (Figure 7.1). The total area of suitable flat soil is 254 hectares, with 29 hectares still exhibiting signs of raised fields.[34] The morphology of these features are similar to the other sites.

---

27 Myers, "Evidence."

28 María del Carmen Molestina, personal communication.

29 "Comisión en Forma a Luis de Castilejo para que Vaya al Pueblo de Carangue Jurisdicción de la Villa de Ibarra y Restituya a Andrés Dias Francisco Ulcoango y Otros Yndios del Pueblo de Carangue de las Tierras que Han Sido Despoxados...," (1635), Archivo Histórico del Banco Central, Ibarra, Juicios, Paquete s/n (Transcribed by Galo Ramón Valarezo).

30 Caillavet, "Toponimia," 12–13.

31 Cieza de León, *Crónica*, 123.

32 Anonymous, "Cibdad," 214–215.

33 See the documents assembled in Tobar Subía, *Monografía*, 1–44.

34 Gondard and López, *Inventario Arqueológico*, 148.

A 240-hectare flat to the north of Otavalo at Pinsaquí (2,560 meters) also exhibits the remains of a few raised fields.

A few other small flats include evidence of raised fields. The site of Sigsicunga at 3,150 meters was examined in the field. Perhaps because of the high elevation, this was the only site where fills between the raised fields had produced organic soils (Histosols). Gondard and López have reported four other small sites to the west of Sigsicunga at elevations between 3,050 and 3,150 meters, as well as a small area at San José de Minas at 2,380 meters.[35]

Functions of Raised Fields

Today, the highland Ecuadorian flats are no longer farmed using raised field methods. The closest analogue may be the double-cropped drained fields which are used for potato and maize cultivation; this highly productive technology has been described above.

Closer raised-field analogues have to be sought elsewhere. In the early 1950s, for example, raised fields were being used to grow vegetables on the altiplano between the city of Bogotá, Colombia, and the National University campus. Elevation and soil were not dissimilar to the Ecuador flats with raised fields. A description of this technology, based on an interview with the late Robert Eidt, therefore may be of value to indicate raised-field possibilities in Ecuador.[36]

The land was a former wetland which was filled with city garbage at the request of about 200 market gardeners. Excess garbage was burned, and the farmers worked the ashes and sawdust into the heavy soils.

The farmers then constructed ridges or platforms 1–2 meters wide. The ditches between the platforms were about 1 meter wide, so that the wavelength was 2–3 meters. The amplitude was about 50 centimeters. The platforms were arranged in an alternating checkerboard pattern. In some areas, water was led into the ditches via canals from rivers. In others, standing water persisted in ditches after the ends of the rainy seasons, allowing splash irrigation using a long pole with a scoop at the end.

The soils were continually reworked by adding garbage, ashes, and sawdust. Once a year the ditches were cleaned out and the muck heaped up on the platform tops. The entire family participated in this task using shovels.

---

35 Gondard and López, *Inventario Arqueológico*, 148.

36 This section is based on studies by Robert Eidt in 1950 and 1951. I interviewed Professor Eidt on this matter before his untimely death; only part of what appears here has been published before. See Eidt, "Aboriginal Chibcha Settlements," 386; and Robert Eidt, "Rural Society and Land Use Change in the Highland Basins of Colombia," *Latin American Studies (University of Tsukuba, Ibaraki Japan)*, 3 (1981), 37–39.

The platforms were double-cropped so that crops were growing throughout the year. Corn, squash, potatoes, and a variety of market garden crops were planted. Frost was a real problem but did not prevent cultivation. Other hazards included flooding in wet years, drying up of streams in dry years, insect pests, bird predation of seeds, and theft. Today, this type of hand agriculture is probably extinct on the Bogotá altiplano. There are modern ditched fields for the production of vegetables, but the ditches are much shallower than before and are probably made with machinery.[37]

Even closer parallels to the probable functioning of the Ecuadorian raised fields may be found in New Guinea. Recent farmers, using digging sticks 1.5 to 1.8 meters long and about 10 centimeters thick, trimmed to a chisel point at one end, dug closely spaced ditches "six or more feet deep and four to six feet wide" in alluvial plains. The purpose of the the ditches was not primarily drainage but rather to:

> get at the rich black swamp deposits and virgin alluvial materials of subsurface levels, which, when spread over the the impoverished topsoil, bring a new lease of life to the land ... the process, as observed, is first to cover the ground with a mattress of cut grass, then to heap the excavated materials on this in a bed 12 to 15 inches thick.

Management of water levels in the ditches was made possible by the construction of small temporary dams in the ditches.[38]

Raised Field Frost Control

To test the hypothesis that prehistoric raised fields performed the function of imiting frost risk,[39] a field experiment was set up at the edge of the Cayambe plain in August and December 1981. This was the first test of raised field frost functioning in the Andes. The results, which have been

---

37 Much smaller scale ridges with standing water in the furrows are used for broad bean production in wet areas of the Vilcanota valley of Peru. These are called simply "wachukuna."

38 See L. J. Brass, "Stone Age Agriculture in New Guinea," *Geographical Review*, 31 (1941), 555–569. Other accounts corroborate this picture, e.g. Karl G. Heider, *The Dugum Dani: A Papuan Culture in the Highlands of West New Guinea*, New York: Viking Fund Publications in Anthropology 49, Wenner-Gren Foundation for Anthropological Research, Inc., 1970.

39 See Eric Waddell, *The Mound Builders*, Seattle: University of Washington Press, 1972; and Thomas Riley and Glen Freimuth, "Field Systems and Frost Drainage in the Prehistoric Agriculture of the Upper Midwest," *American Antiquity*, 44 (1979), 271–285.

published elsewhere,[40] are summarized here. Subsequent tests of raised field frost functioning in Peru have tended to corroborate these findings.[41]

Three thermometers were placed at the test site; one on an unridged flat, one on a 4 percent slope, and one on a ditch bank in the flats. At the latter site, a tank and connecting ditch were dug to mimic prehistoric raised field cross-sectional characteristics.[42]

Minimum nightly temperatures were read for two one-week periods, at the end of the dry season and again in the middle of the rainy season. The results showed that flat minima usually were colder than slope minima. On cold nights, the difference between slopes and flats is accentuated, presumably because cold nights tend to be windless and cloudless and offer the greatest opportunity for cold air drainage and radiative cooling to create inversion layers.

The thermometer on the simulated raised field always showed an improvement over the adjacent flat. This was true even on warm nights when the slopes were cooler than the flats. The effect seems to be due not to cold air drainage but rather to the formation of tubes of warm air over the ditches with standing water; these tubes of air then warm the adjacent ridge. A linear regression between temperatures on the raised field and on the adjacent flat suggests that the ridge will experience a frost of 0 degrees only when the adjacent flat has temperatures of –1.3 degrees. This improvement is sufficient in this environment to produce a lengthening of the growing season by months.

The effect disappears when one of the two ditches flanking the raised field is filled in. Closely-spaced ditches in highland raised fields function, then, to ameliorate frost risk, and this may explain why tropical lowland raised fields tend to have larger wavelengths.

Raised Field Fertility Control

The soil quality of prehistoric raised fields depends on the agricultural practices involved in their use. If frequent ditch cleaning was necessary to maintain ditch depth at over 0.8 meters, the ditch spoil itself (placed on the ridge tops) is the material which determines the initial soil quality of raised field surfaces.

---

40 See Knapp and Denevan, "Use of Wetlands;" Knapp, *Soil.*

41 See the application of a heat storage model to raised fields in Alan Kolata and Charles Ortloff, "Thermal Analysis of Tiwanaku Raised Field Systems in the Lake Titicaca Basin of Bolivia," *Journal of Archaeological Science*, 16 (1989), 233–263.

42 All three thermometers were placed far from walls or trees, on surfaces covered with dense kikuyu grass. The thermometer bulbs were all 30 centimeters above the surface and at least 10 centimeters above the grass layer. The various thermometers were calibrated against each other and against a reference thermometer in the meteorological station shelter at San Pablo; all readings were converted to San Pablo shelter equivalents.

Soil samples were collected from a variety of ditches between raised fields; some of these are shown in Table 3.3.[43] Low nutrient status was characteristic of soils in (1) low pH environments such as Sigsicunga or La Tola (Cayambe), or (2) volcanic ash falls. e.g. at Chillogallo. If low pH and volcanic ash deposits are eliminated, the average phosphate values for raised field ditch fills is 107 kg/hectare. This can be favorably compared with the average of 127 kg/ha for accumulated silt in irrigation tanks in the Pimampiro area (Table 3.3), especially since the raised field ditch fill is much more available for spreading on field surfaces. In contrast, the quality of soils on relic raised field surfaces is much less; about 25 kilograms/hectare phosphate.[44] Why is the ditch fill richer in phosphate? It is likely that part of the answer has to do with the the accumulation of available phosphorus which has been leached from drainage basin soils and brought into ditches. It is also possible that the organic matter in the ditches may supply some phosphorus as it decays.[45]

Raised Field Water Control

The water table of the San Pablo flat was periodically measured in an abandoned well though the period December 1980 – September 1981. Very little seasonal fluctuation was observed; water levels fluctuated between 50 and 70 centimeters of the field surface, with minima during the summer dry season (June–September) and the little summer (January).[46] The maintenance of very widely spaced drainage ditches (200 meters apart) seems adequate to prevent flooding.

In this environment, ditches between raised fields probably would always have standing water, especially given the checkerboard pattern of fields and the availability of perennial flow from the nearby Itambi River.

In the Cayambe flats, the situation is somewhat different. Here the water table is typically over one meter below the surface, and the river has entrenched itself in the contemporary plain. The unique semi-circular embankments of this flat therefore make sense as water conservation devices. Farmers would have been favored by the unusual flow pattern of the Granobles River, which actually peaks in the summer dry season.[47]

---

43 The full sample set is shown in Knapp, *Ecología*, 153.

44 For those soils with pH between 6 and 7.

45 I was fortunate to be able to discuss these issues with the late Robert Eidt, whose insights are incorporated here.

46 See graph in Knapp, *Ecología*, 158.

47 PRONAREG-ORSTOM, *Estudio Hidro-Meteorológico*, 114. See also Batchelor, "Camellones." The local hacienda owner Marcos Wray Espinosa, who has also studied these fields, informed me that the features make no sense as drainage devices given the drought hazard here, but rather can best be understood as facilitating irrigation.

The other sites with raised fields all have high water tables and/or are near perennial streams or springs. Irrigation would not have been difficult. Irrigation, coupled with frost control, would have enabled double cropping.

Estimating Raised Field Productivity

Raised fields modify frost risk and drought risk and thus enable double cropping. Contemporary ethnographic analogies suggest that potato-maize double cropping is a possibility, and sixteenth century sources are consistent with this interpretation if not conclusive on the subject.

Productivity in terms of total area would be reduced by the fact that so much land is taken up by drainage ditches. No known New World crop is hydromorphic, and even if totora reeds were grown in ditches, they would not contribute much to group calorie intake. The wavelength of highland raised fields is small—about 3 to 7 meters—and it is likely that up to half the land surface was uncultivable.

Yields of highland raised fields would have depended on fertility. This would have been augmented by the annual or twice annual mucking process, spreading material with a phosphate content of perhaps 107 kilograms per hectare on field surfaces. Ditch cleaning at Hacienda La Vega produces 50 cubic meters of spoil per 100 linear meters of ditch. If we assume that raised field ditches were spaced 4.5 meters apart, there would have been at least 2,220 linear meters of ditch per total hectare of land. Cleaning of these ditches would produce 1,110 cubic meters of wet spoil. Spread over one half hectare of cropped raised field tops, this would have totally transformed the upper 22 centimeters of topsoil ("the plow layer") to resemble muck fertility. The potential yields will be modelled in the next chapter.

Costs of Raised Fields

Raised fields obviously would have involved considerable labor. The problem is not so much that of constructing the fields; the fields could have been constructed slowly and incrementally by individual families.[48]

---

[48] See William Doolittle, "Agricultural Change as an Incremental Process," *Annals of the Association of American Geographers*, 74 (1984), 124–137. Erickson goes farther by suggesting that southern Peruvian raised fields not only could be constructed incrementally but that the ditches do not require periodic mucking and that fields yield continuously without annual fertilization or replenishment from ditch fills; thus he considers the technology to be labor-saving. It will be interesting to see the long-run ability of raised fields to maintain good yields; the fields which have been tested for two consecutive crops produced 20 percent fewer tons/hectare of potatoes during the second year (but n = 2); see Erickson, *Archaeological Investigation*, 244.

Rather, the problem would have been the mucking process by which the ditches were cleaned and muck deposited on the field surfaces.

At present-day work rates, these ditches would have required 7.5 person-days per 100 meters for cleaning—166.5 person-days per hectare. If, as assumed, pre-Hispanic tasks required 2.7 times as much time to complete, 450 person days per total hectare and 900 person days per cultivated hectare would have been required. These values are extraordinarily high. Are they comparable to competitive technologies at high population densities? The modelling approach employed in the next chapter will indicate a possible answer to this question.

Two High-Productivity Systems

This chapter and the previous chapter have provided—in considerable detail—evidence that at least two high-productivity agricultural technologies were deployed in late pre-Hispanic times. Similar in terms of land efficiency, they were very different in labor efficiency and in geographical "reach." Irrigation used the power of moving water to reduce drought risk, enable cropping in regions of minimal risk of downpour, and provide nutrients to soil. It required the construction and (not particularly onerous) maintenance of canals. The canals in turn were vulnerable to variety of environmental and social risks.

Raised fields did not require the construction of vulnerable long distance agricultural landforms, and indeed probably insulated societies from risk rather than exposing them to greater risk. But the labor of water lifting and mucking is enormous, so that the high yields of raised field farming are probably only justifiable at high levels of demand.

Both systems, along with rainfed agriculture, were demonstrably in use in pre-Hispanic times. In the following chapter, a modelling approach will be used to indicate and test hypotheses about their relative contribution to pre-Hispanic subsistence.

Chapter 8

ADAPTATION AND SETTLEMENT IN LATE PREHISTORIC TIMES

The materials in the previous three chapters suggest substantial differences between adaptive options in late pre-Hispanic times and the present. These can be incorporated into the modelling approach employed earlier to suggest labor and land efficiencies of different technologies in different niches, which in turn can be incorporated in scenarios of adaptation.

The size of the ordinary pre-Hispanic farming household is difficult to estimate. As noted previously, the census units of the visita analyzed by Salomon averaged about 4.06 persons, but included smaller units which were not de facto households. We will assume that the "typical" household included five persons, basically a married couple and children.

The number of harvest calories needed to support a person per year (d), would have been greater in pre-Hispanic times. Accepting the contemporary highlander consumption need of 1,733 calories per day,[1] a minimum of 0.63 million calories per year would have been required for subsistence. Potato or maize production needed to be higher than this, however, because of loss to pests in storage. Available data suggest that post-harvest losses of the order of 25 to 30 percent may be typical of subsistence farms in Latin America.[2] Thus a production of a million calories per person may be considered sufficient; for subsistence needs. The goals of the pre-Hispanic household were both subsistence and the production of festive gifts. For the purposes of modelling, it will be assumed that maize-producing households strove to produce 20 percent more than the subsistence need (or 1.2 million calories per person per year),

---

1 The average calorie consumption of all individuals above 1 year of age in three villages in northern Pichincha province, north of Quito, was 1,769 calories per day; for Peguche, near Otavalo, the value was 1,697 calories per day. Averaging these two values gives 1,733 calories per person. See Vargas and Gallegos, *Diagnóstico*.

2 See NRC (Commission on International Relations, National Resource Council Board on Science and Technology for International Development), *Postharvest Food Losses in Developing Countries*, Washington D.C.: National Academy of Sciences, 1978.

in order to produce maize beer, where the dregs from manufacture were fed to guinea pigs. It will be assumed that potato producing households produced 40 percent more than the subsistence necessity (or 1.4 million calories per person per year), to compensate for the lower festive significance of potatoes.[3]

Modelling Agricultural Intensification in Ordinary Farms

As discussed in Chapter 5, it will be assumed here that all tasks except maize harvest and irrigation took 2.7 times as long before the advent of metal tools; it will be recalled that this value was based on New Guinea values for grassland reclamation and ditch digging with and without metal tools.

In prehistoric times, manure resources would have been more restricted than at present. The household manure resource was primarily that available from humans and guinea pigs—perhaps about 1,140 kilograms (dry weight equivalent) manure per family per year—equivalent to 33 kilograms nitrogen and 16.1 kilograms phosphate. If 40 percent of the nitrogen and 60 percent of the phosphate were available, 13.2 kilograms nitrogen and 9.7 kilograms phosphate would have been delivered to agricultural fields. There would have been little difficulty in gathering such manure and the transport of manure to the fields could have been accomplished during trips to the fields for other purposes; little labor time would have been added by manuring. We will assume however (and basically without historical evidence) that prehistoric farmers applied all the family and guinea pig manure to their fields.

A household manure resource of 1,140 kilograms would not have been sufficient of itself to enable agriculture without other sources of soil nutrients. Indeed a family of five, requiring 5 million calories a year, would consume 1,385 kilograms of maize;[4] the maize plants, however, would use 36 kilograms of nitrogen to produce this quantity of calories.[5] The family manure resource produces only 33 kilograms of nitrogen (13.2 kilograms immediately available). Obviously, then, families would have required nutrient release from soil organic matter, or some supplemental fertility from irrigation or mucking, to feed themselves in prehistoric times.

---

3 The ethnohistoric evidence for lavish gifts of maize beer can however be interpreted as a means of eliminating the surplus in better than average years, without needing to assume a greater than subsistence production in normal years. The greater production for potatoes can be interpreted as expended in trade, reciprocal gifts, or as representing the labor cost of cultivating extra fields in the maize zone.

4 Leung, *Tabla*.

5 Donohue, Schickluna and Robertson, *Soils*, 320.

If the modelling approach described previously is used given the above assumptions, then unirrigated maize and potatoes can be expected to respond accordingly (Tables 8.1 and 8.2).

TABLE 8.1 MODELLED PREHISTORIC POTATO YIELDS AND LABOR

Overall population density (persons/km.^{2})	Fallow (years)	Yield (tons/ha.)	Labor (person days/ family)	Manure (tons/ha.)
10	57	11.2	385	1.6
20	21	8.7	440	0.9
30	10	7.0	508	0.5
40	5.5	5.5	614	0.3
50	*Not possible to support this population density*			

TABLE 8.2 MODELLED PREHISTORIC MAIZE YIELDS AND LABOR

Overall population density (persons/km.^{2})	Fallow (years)	Yield (tons/ha.)	Labor (person-days /family)	Manure (tons/ha.)
20	28	2.1	169	1.5
30	15	1.9	179	1.1
40	9.4	1.6	189	0.8
50	6.1	1.5	202	0.6
60	3.9	1.3	217	0.4
70	*Not possible to support this population density*			

Irrigation and Raised Fields: Ordinary Farms

Irrigation in pre-Hispanic times would have involved roughly the same adjustments as for unirrigated maize: higher labor inputs (except for irrigation itself) and lower fertilizer applications. The modelled results are shown in Table 8.3.

Raised fields are a special case. We will assume ditch cleaning once a year (at a cost of 450 person-days per total hectare); no reclamation labor; an initial fertility corresponding with that of the muck in ditches (107 kg/ha phosphate); and initial nitrogen corresponding with that of moderately fallowed fields (0 kg/ha using the fertility index equation). The modelled results are shown in Table 8.4.

TABLE 8.3 PREHISTORIC IRRIGATED MAIZE YIELDS AND LABOR*

Overall population density (persons/km.^{2})	Fallow (years)	Yield (tons/ha.)	Labor (person-days/ family)	Manure (tons/ha.)
50	17.7	3.1	142	4.1
100	6.2	2.4	164	2.4
150	2.7	1.9	193	1.4
160	2.3	1.8	203	1.3
170	*Not possible to support this population density*			

*Assuming 700 millimeters a year mean annual precipitation.

TABLE 8.4 PREHISTORIC RAISED FIELD YIELDS AND LABOR: ORDINARY FARMS

Overall population density (persons/km.^{2})	Fallow (years)	Yield (tons/ha.)	Labor (person-days/ family)	Manure (tons/ha.)
100	8.5	15.4	697	2.1
300	2.2	15.4	697	2.1
500	0.7	15.4	697	2.1
700	0.35	15.4	697	2.1
750	0.26	15.4	697	2.1
800	0.18	15.4	697	2.1
900	*Not possible to support this population density*			

Tables 8.5 and 8.6 show the population densities of various niches, assuming that labor inputs are equalized across space. Under conditions of equal labor allocation, it can be expected that (1) maize areas would have supported denser populations and lower fallow times than potato areas; and thus (2) the lower elevations suitable for maize production would not have been employed for potato production, and (3) populations would have dropped off away from the optimum precipitation value.

Furthermore, Table 8.3 suggests that irrigated areas would have had very high population densities even under conditions of low labor input. Table 8.4 suggests that under low prevailing population densities and labor inputs, raised field areas would not be used, while under conditions of high population densities raised field areas would be more densely populated than any other.

TABLE 8.5 PREHISTORIC POPULATION DENSITIES BY CROP, ASSUMING COMPARABLE LABOR EFFICIENCIES

MEDIAN PRECIPITATION	YIELD CONSTANTS, MAIZE AND POTATOES	Calculated population densities (persons/km.^{2}): *Marginal labor input = 385 person-days/family**		*Marginal labor input = 700 person-days/family***	
		MAIZE	POTATOES	MAIZE	POTATOES
82	5/22	90	20	90	55
94	4/19.6	60	10	60	44
106	3/17.2	32	0	46	32
118	2/14.7	0	0	23	19
130	1/12.3	0	0	0	9
Average, 94,106, and 118 mm		31	3	43	32

*Irrigated maize would support a density of 160 persons per square kilometer at this labor input level and at 700 millimeters mean annual precipitation.

**At this labor input, irrigation would support a density of 160 persons per square kilometer at 700 millimeters mean annual precipitation, while raised fields would support 800.persons per square kilometer.

TABLE 8.6 PREHISTORIC POPULATION DENSITIES BY CROP, ASSUMING COMPARABLE LABOR EFFICIENCIES: IRRIGATION AND RAISED FIELDS

MEDIAN PRECIPITATION (MM./MONTH)	Calculated population densities per square kilometer*: *Marginal labor input = 385 person-days/family*		*Marginal labor input = 700 person-days/family*	
	IRRIGATED MAIZE	RAISED FIELDS	IRRIGATED MAIZE	RAISED FIELDS
58	200	1450	200	1450
70	150	0	150	1200
82	110	0	110	1000
94	75	0	80	800
106	25	0	55	0
118	0	0	20	0
130	0	0	0	0

*Rounded to nearest 10 for values over 100 and nearest 5 for values under 100.

Chief's Lands

The logic of chief's land use is that of ordinary land use writ large: subsistence plus the production of a surplus usable for festival purposes. The number of people in the chief's household was larger, so that a larger amount of people had to be fed; but presumably the increase in mouths was matched by an equivalent increase in manure supply from humans and guinea pigs.

If manure availability is proportional to population, the larger size of the chief's household should have no effect on the relative modelled yields and labor efficiencies of the various options.

Some differences may have been apparent in manure supply and in production however. If the larger household had more guinea pigs than average, and/or if the chief's camelid herd's dung was available for fertilization, fertilization rates on the chief's fields may have been higher, resulting less work to feed equivalent populations. If a greater proportion of the chief's produce went to non-subsistence purposes (calorifically inefficient guinea pig or maize beer production), this effect would have been counterbalanced, and the amount of work to feed a given population may have been comparable.

More important, perhaps, is the difference in labor supply. Chiefs had access to servile labor, which perhaps could provide more work per family per year than for the population as a whole. More importantly, chiefs had access to corvée labor. It is easier to muster large amounts of corvée labor once a year for a simple repetitive task than to muster small amounts of labor numerous times a year for different complex tasks. Corvée labor operates best in gangs performing slow and easily supervised labor in a geographically restricted area. These characteristics of simplicity, slowness, geographic concentration, and seasonal concentration are especially characteristic of the tasks of ditch cleaning, whether for canal irrigation or for raised field mucking.

If corvée labor was primarily available for ditch cleaning and mucking, the form of agriculture which would most benefit was raised field farming. Fully 68 percent of the modelled labor in raised field farming was due to mucking; and if much of the mucking were provided by corvée labor, the remaining labor for raised fields would decline greatly. It is noteworthy that corvée labor would have provided relatively little benefit to irrigation, since modelled canal cleaning labor was only about a tenth of total labor requirements.

Therefore, the chief's economic unit in general probably had the same preferences as a small household in terms of agricultural options. With corvée labor, the relative attractiveness of raised fields would have been increased, if not to the point of being fully competitive with irrigation, at least to the point of being highly competitive with unirrigated maize and potato farming.

A Pre-Hispanic Scenario

With the modelled yields and labor efficiencies provided here, it is possible to generate internally consistent scenarios of pre-Hispanic land use (see Tables 8.5 and 8.6). Given a series of assumptions (most notably an assumption of relative equality of per capita labor allocation across households), and assuming that raised fields were used (at 700 person-days per family), the population supported by various strategies would have varied greatly across the landscape; at typical precipitation (94 mm/month, 1,030 mm/year), 44 persons per square kilometer would have been supported by potato farming, 60 to 80 persons per square kilometer would have been supported by maize farming, depending on whether irrigation were employed; while 800 or so could have been supported per square kilometer by raised fields. The approach employed here suggests further that population would have rapidly declined at higher precipitation values, so that at precipitation values higher than 118 mm/month (1,293 mm/year) agriculture would not be practiced. At lower precipitation levels, both raised fields and irrigated fields would have supported higher populations. Elevation would not be correlated with persons supported, except at the upper limit of maize cultivation where potato farming would have resulted in lower densities.

The most attractive (labor efficient) niche would have been maize irrigation, with raised fields perhaps being competitive in the chief's economic unit where corvee labor was available.

If corvée labor brought the effective cost to the chief of raised fields down to 385 person-days, for example, at 94 millimeters per month unirrigated maize farming would still support 60 persons per square kilometer (and irrigated maize farming 75 persons), but potato farming even near optimal precipitation values would have supported only 10 persons per square kilometer. Furthermore, maize farming would now drop out at precipitation values above 106 millimeters/month (1,161 mm/year) (see Tables 8.5 and 8.6).

In the remainder of this chapter the reasonableness of these scenarios will be checked against two independent sets of data: data on settlement location, and data on prehistoric population densities.

Before proceeding to these tests, however, the assumptions underlying this analysis will be emphasized again. These assumptions limit the possibilities of applying these models and scenarios without modification to areas outside the equatorial Andes.

A number of assumptions were made about the physical environment, especially about the nature of the soils. The yield equations were based on experience with fertile volcanic soils, and are therefore not directly applicable to non-volcanic soils. The relationships between precipitation and yields are based on the particular distribution of rainfall found at the Plaza farms, probably representative of much of the equatorial Andes, but not other regions.

The projections for prehistoric yields are based on an assumed scarcity of manure, which however is assumed to have been applied. Higher yields could be expected where camelid manure was relatively abundantly available, as perhaps in the southern Peruvian highlands or even in the vicinity of San Pablo, Ecuador. The possible increases in production associated with interplanting or rotation with leguminous plants are not included, nor for that matter the production of auxiliary crops (quinoa, beans, tarwi, fruits). The inclusion of these items might increase the population estimates somewhat. At the same time, loss to pests may conceivably have been greater in prehistoric times than contemporary experience indicates.

It has been assumed that 1.2 million calories supported an average person for a year. The implicit assumption of over half of calories to pests, spoilage, and calorifically inefficient uses (beer for festivals) may need to altered as knowledge improves.

It will be assumed, as discussed earlier, that most people lived scattered about the countryside in reasonably close proximity to their fields. Thus population patterns mirrored production patterns, and land use intensity was not notably affected by differences in accessibility from population centers. A partial exception might be the chief's fields, where the larger population of the chief's household in effect created a somewhat nucleated settlement with a corresponding tendency to increase intensification in the vicinity of the tolas.

It will be assumed that relatively little land was unused due to location midway between adjacent llajtakuna (e.g. for military reasons).

More importantly, the scenario generally involves the assumption that labor inputs were roughly equal between families, irrespective of niche and technique. As previously discussed, this hypothesis of least effort is in fact consistent with a variety of conscious and unconscious processes, and a variety of modes of production. In particular, where subsistence coupled with surplus production for festival use is the rationale of production, it can be expected that labor efficient technologies and niches will be "selected for."

Thus it is assumed that land use in the equatorial Andes can be partly understood or explained in terms of the labor and land efficiencies of competing technical options of production, where the strategically relevant environment is contingent on prevailing population density, socio-economic organization, political structure, as well as the prevailing tool kits of producing and preparing foods.

Hypotheses and Statistical Tests

Since for the most part, common people lived in houses which are dispersed and difficult to locate, archaeology cannot be used to directly reconstruct population distributions and densities for this segment of the population. The chief's residence sites have been preserved, however, in the tola complexes, and it has been argued that the chiefs located their houses in the midst of their agricultural fields.

The locations of tola sites, therefore, can be used as a test of the major conclusions of the scenarios of pre-Hispanic land use developed above. In particular, if the scenario has value, ramp tola sites should be concentrated (1) in flats suitable for raised field cultivation or irrigation, as opposed to slopes; (2) in lower elevations suitable for maize cultivation, rather than higher elevations; and (3) in areas with moderate precipitation rather than in areas with very high or very low precipitation. Histograms of the number of ramp tola sites plotted against elevation should show little influence of elevation on settlement density up to the upper limit of maize cultivation, but a marked decrease of settlement density at higher precipitation values. Depending on the scenario chosen, settlement should vanish at precipitation values above 1,160–1,290 millimeters a year.

Tolas are found in the highland area roughly from Quito to the Mira River. Most of the known sites with ramp tolas are shown in Figure 8.1 and Table 8.7. Ramp tolas were built from about A.D. 1250 to about A.D. 1525; excavations at the Cochasquí and Socapamba sites indicate that the tolas were continuously occupied.[6] Thus the sites displayed in Figure 8.1 probably represent for the most part simultaneous rather than successive chiefly house sites.

It is possible that in some cases chiefs' houses were not located on top of ramp tolas. However, since the building material for tolas was ubiquitous in the region, and since tolas were built in both slopes and flats, it is hard to see that any bias in the sample would be introduced by this factor.

In some cases ramp tolas may have been destroyed. If one is right in believing that tolas are located in or near agriculturally desirable areas, there would have been an incentive to destroy them for field expansion or town building. The ramp tolas are so large, however, that it is hard to believe that more than a few have been completely destroyed. Partial destruction has probably been more usual. The tolas at Otavalo, for example, have been sufficiently disturbed that it is no longer possible to tell if they once had ramps.[7] An immense tola to the east of Cayambe apparently once had a ramp.[8]

---

6 Athens, *El Proceso Evolutivo*, 126, 133, 135–136; Albert Meyers, "Análisis de la Cerámica de Cochasquí," in Udo Oberem (ed.), *Cochasquí: Estudios Arqueológicos*, Otavalo: Instituto Otavaleño de Antropología, 1981, 265.

7 Athens, *El Proceso Evolutivo*, 237.

8 Jijón y Caamaño, *Contribución*, 293–294.

TABLE 8.7 RAMP TOLA SITES

Site	Annual Precip. (mm.)	Elevation (meters)	Distance to flats (km.)	Irrigated?	Raised fields?	Gondard and López site
1	900	2450	4	Yes	—	I-42
2	800	2450	0	Yes	—	39
3	900	2350	2.4	—	—	37
4	1000	2700	5.9	—	—	24
5	600	2200	3	Yes	—	58
6	800	2200	0	Possibly	—	66
7	800	2450	0.8	Yes	—	112
8	950	2700	4	Yes	—	108
9	800	2550	2	Yes	—	109
10	1000	2550	1.7	Yes	Yes	105
11	1200	2800	0.5	—	Yes	99
12	1100	2750	0	—	Yes	93
13	1100	2900	0	—	—	89
14	1150	2900	0	—	—	90
15	800	2950	0	—	Yes	P-89
16	800	2850	0.5	—	Yes	94
17	750	2800	0	—	Yes	93
18	1000	2950	8	—	—	77
19	950	2300	4	—	Yes	4
20	1100	2500	0	—	—	42
21	500	2100	2.5	Yes	—	38
22	800	2300	1	—	Yes	I-82

The area shown in Figure 8.1 was probably entirely known to the builders of ramp tolas. Ramp tola sites are present to the north, west, east, and south of the area shown.[9] In the figure are indicated the locations of all ramp tola sites reported by the surveys of Gondard and López and Athens, plus the Cayambe ramp tola reported by Jijón y Caamaño. In addition, the Caranque site (number 22 in Figure 8.1) is indicated, as it is apparent that this site also had at least one ramp tola, with the ramp being destroyed by brickmaking operations.

It is expected, if the scenario has validity, that a significantly greater number of ramp tola sites should occur in certain kinds of environments than if location were completely random. Ramp tola sites should be

9 Sites I-071, I-001, C-113, I-071, I-016 in Gondard and López, *Inventario Arqueológico*; site Pi-10 in Athens, *El Proceso Evolutivo*, 268.

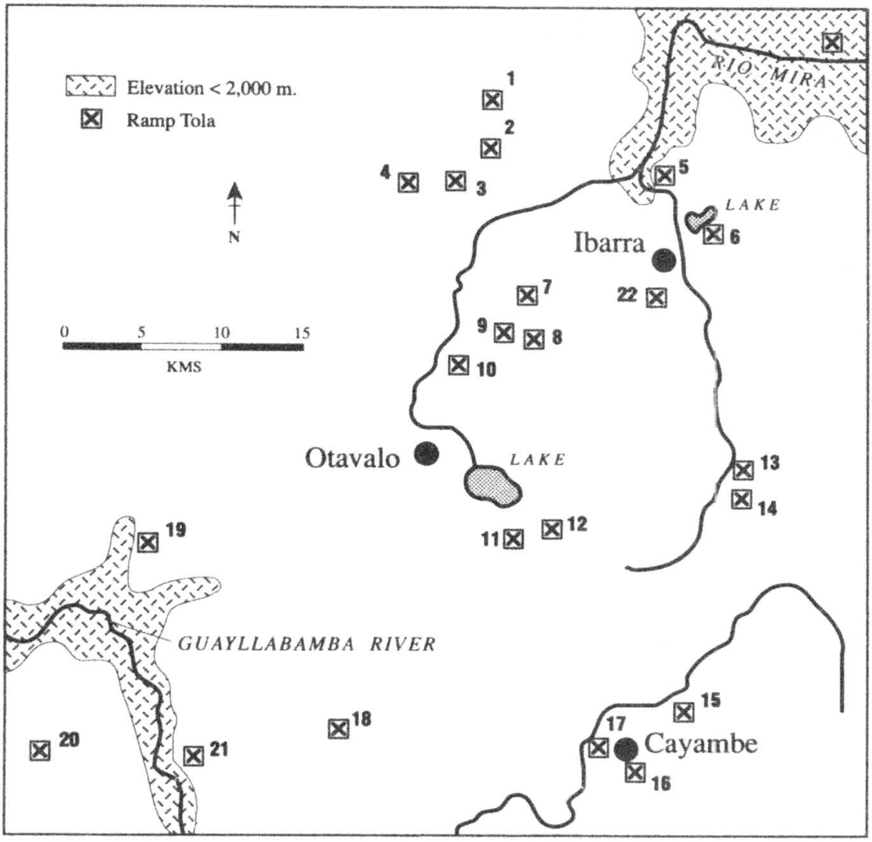

Figure 8.1 Location of Ramp Tola Sites, Northern Highland Study Region

concentrated in flats, in elevations where maize can be grown, and in areas without excessively high precipitation, other things being equal.

As phrased, these questions refer to comparing the frequency of events in intervals (special areas) in which the expected frequency for the null hypothesis (randomness of distribution) can be calculated as the proportion between the special areas and the total area. The probability of the observed frequency being the result of random processes can be calculated using the binomial distribution.[10]

---

10 Jim Hodges first suggested to me the importance of this distribution with respect to this problem. See the discussion in Herbert Arkin and Raymond Colton, *Statistical Methods*, New York: Barnes and Noble, 1970, 121–122. The binomial distribution can be readily calculated with the Excel spreadsheet program on a Macintosh computer.

Two special problems present themselves in calculating the probability of the observed distribution of ramp tola sites. First, the results are dependent on the area chosen for study. The boundaries of Figure 8.1 roughly correspond on three sides to natural boundaries: the Andean crest to the east, the canyon of the Mira River to the north, and the canyon of the Guayllabamba River to the south. The western boundary does not coincide even roughly with any obvious boundary in the natural landscape; it could not be extended much farther west, however, without leaving the area of known ramp tola sites. Second, given the small number of ramp tola sites, it is difficult to control for the environmental factors not under immediate consideration in each test (slope, elevation, precipitation). These factors will be controlled for where possible.

In Table 8.7, precipitation totals for each site have been determined from values for the 1964–1973 decade. This may be a fair representation of conditions during late prehistoric and colonial times.[11] Elevations were measured with a dot grid from IGM topographic maps. The areas of non-sandy flats have been measured with a planimeter on the relevant soil map sheets.[12]

From the area on Figure 8.1 the land above 3,500 meters and below 2,000 meters will be excluded from further consideration. In Table 8.8 land area in two elevation categories and three precipitation categories have been listed.[13] In Table 8.9, the area of non-sandy flats have been listed for for each of the environmental categories.

It is now possible to proceed to a discussion of the hypotheses.

Concentration of Ramp Tolas in Flats

Most of the flats are in the area with 600–1,300 millimeters annual precipitation and with elevations between 2,000 and 3,000 meters. Within this area of 1,070 square kilometers, 97 square kilometers are non-sandy flats. However, 8 of the 22 ramp tola sites in the larger area are located in flats. Use of the binomial distribution function indicates that the probability of this concentration occurring only due to chance is under 0.01 percent. We conclude that there is a real tendency for ramp tola sites to be located in or next to flats. This is consistent with expectations (but of course does not "prove") that ramp tola site were located in the most productive land and that the flats were most productive.

---

11 See PRONAREG-ORSTOM, *Isoyetas*. During the 1964–1973 period in Quito, there were 3 GA winters and no GB winters, and average December–April precipitation was only 88 percent of the 1891–1963 average.

12 PRONAREG-ORSTOM, *Mapas de Suelos*.

13 The category boundaries were determined partly by the nature of the hypotheses being tested and partly by the intervals used on available small-scale maps.

TABLE 8.8 ENVIRONMENTAL CHARACTERISTICS AND AREAS, TEMPERATE ELEVATIONS (2,000–3,500 METERS), NORTHERN HIGHLAND STUDY AREA

Precipitation	Area in elevation range (km.^{2})	
(MM.)	2,000–3,000 METERS	3,000–3,500 METERS
under 600	115	0
600–1,300	1,070*	230
over 1,300	240	390

*Of which 670 is over 2500 meters.

TABLE 8.9 AREAS IN FLATS, BY ENVIRONMENTAL CHARACTERISTICS, NORTHERN HIGHLAND STUDY AREA

Precipitation	Area (km.^{2}) in elevation range in flats (except sandy soils)*	
(MM.)	2,000–3,000 METERS	3,000–3,500 METERS
under 600	1.2	0
600–1,300	97	5.8
over 1,300	0	1.2

*Sandy soils are Vitrandepts and Psamments.

Concentration of Ramp Tolas in Lower Elevations

Again, limiting the analysis to areas with 600–1,300 millimeters annual precipitation, there are 1,300 square kilometers total area, of which 1,070 square kilometers or 82.3 percent are at or below 3,000 meters elevation (roughly, the maize elevations). Of the 22 ramp tola sites in the precipitation interval, 100 percent are located at or under 3,000 meters in the maize elevations. the binomial distribution function indicates that the chances of this concentration occurring by chance are only 1.4 percent.
If flats are eliminated from consideration, the total area is 1,197 square kilometers, 81.3 percent of which are below 3,000 meters; of the 14 ramp tola sites outside of the flats, 100 percent are in the maize elevations. The binomial distribution function indicates indicates that the probability of such a distribution occurring by chance is 5.5 percent. The sample size, however, is small, and the results are still encouraging in view of the hypothesis.

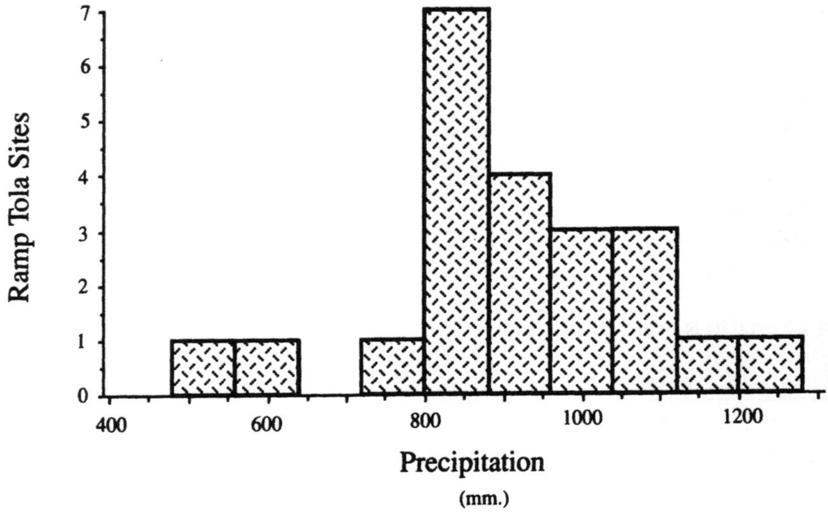

Figure 8.2 Histogram of Ramp Tola Sites by Precipitation Category

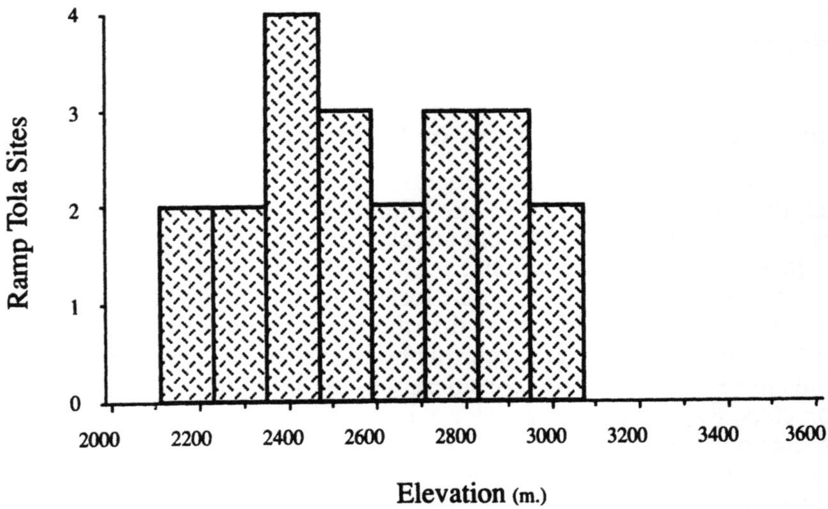

Figure 8.3 Histogram of Ramp Tola Sites by Elevation Category

Concentration of Ramp Tolas in Areas with Moderate Precipitation

If the area between 2,000 and 3,000 meters alone is examined, there are 1,425 square kilometers, of which 1,185 or 83 percent experience less than 1,300 millimeters per year precipitation. Of the 22 ramp tolas in the elevation interval, 100 percent occur in areas with less than 1,300 millimeters precipitation. The bimodal distribution function indicates that the probability of this concentration occurring by chance is only 1.7 percent. This is consistent with the notion that agricultural production is maximal below 1,300 millimeters a year annual precipitation.

If the levels are eliminated from consideration, there remain 1,327 square kilometers, of which 1,087 or 82 percent receive less than 1,300 millimeters rainfall per year; again, 100 percent of the 14 ramp tola sites not on flats are in areas experiencing less than 1,300 millimeters per year. The binomial distribution function indicates that the probability of this distribution occurring by chance is 6.1 percent. As with the case of the potato elevations, we can assume non-random factors are operating with a moderate level of confidence, particularly in view of the small sample size.

Averages and Distributions

The average annual precipitation for all ramp tola sites is 900 millimeters per year. The numbers of ramp tola sites appear to fall off at higher precipitation values (Figure 8.2), and appear slightly more consistent with the low labor input scenario than with the high labor input scenario.

The average elevation of the ramp tola sites is 2,577 meters. The tola distribution with respect to elevation is relatively flat from 2,000 to 3,000 meters (Figure 8.3). This does not cast doubt on the scenarios' prediction of relatively constant yields and populations throughout the maize elevations.

Averages and Distributions: Equatorial Andes

Out of interest in checking the hypotheses about the roles of elevation and precipitation in settlement distribution, 54 town sites were selected in the equatorial Andes known to correspond very roughly to pre-Hispanic llajtakuna.[14] Mean annual precipitation for these sites[15] averaged 893

---

14 See Larrain Barros, *Demografia*, Frank Salomon, *Señores Etnicos*; and John Murra, "The Historic Tribes of Ecuador," in Julian Steward (ed.), *Handbook of South American Indians, Volume 2*, Washington, D.C.: Smithsonian Institution Bureau of American Ethnology, Bulletin 143, 1944, 785–822. The towns are: Tulcán, Huaca, San Gabriel (Tuza), El Angel, Mira, Pimampiro, Cahuasquí, Guayllabamba, El Quinche, Pomasquí, Zambiza/Nayón, Cotocollao, Chillogallo, Calacalí, Nono, Lloa, Cumbaya, Tumbaco, Yaruquí, Puembo, Pifo, Alangasí, El Inca, Pintag,

millimeters per year, similar to that of the ramp tola sites listed above. Only 6 of the towns experienced over 1,300 millimeters per year. It is of interest that local witnesses interviewed in 1559 in one of these "wet" towns noted that maize yields decreased if it rained much.[16]

Mean elevation of the 54 sites is 2,724 meters; only 3 of the 54 towns is located above 3,100 meters.

The distribution of the town sites is consistent with expectations for concentration in the maize elevations and in areas with moderate precipitation.

Pre-Inca Population Densities: Other Evidence

If the distribution of ramp tola sites is consistent with the scenarios of pre-Hispanic land use and population density, it is also revealing to look at the evidence for overall populations in the region under study.

Table 8.5 suggests that populations growing rainfed maize and potatoes varied according to precipitation, decreasing at higher precipitation values. This decrease would have at least been matched on the dry side as droughts took their toll. The average populations of the three wetter trials (94, 106, 118 millimeters per month) might be taken to be not far different from the three dryer trials (58, 70, and 82 millimeters per month), allowing an estimate of overall average population density for the 600–1,300 millimeter interval. For the scenario of 700 overall person-days per family labor input, this average population density would be 43 persons per square kilometer for the maize elevations and 32 persons per square kilometer for the potato elevations. The 230 square kilometers of high slopes in the study region with between 600 and 1,300 millimeters a year precipitation could then have supported 7,360 (230 x 32) persons. Using the same argumentation, the 973 kilometers of lower slopes (other than flats) could have supported 41,839 (43 x 973) persons. The two zones combined would have supported 49,199 persons.

To this value can be added the population supported by irrigation and raised fields. If one assumes that 5 square kilometers were irrigated at Atuntaqui, and 2 square kilometers each devoted to maize production (in addition to tropical specialty production) at Tumbabiro and Urcuquí, about 9 x 160 or 1,440 persons would have been supported at these dry (about 700 mm/year) sites (Table 8.6).

Sangolquí, Amaguaña, Conocoto, Uyumbicho (modern Tambillo), Aloag, Aloasí, Mulalo, Saquisilí, Alaquez, Latacunga, Mulalillo, Ambato, Tisaleo, Quero, Mocha, Píllaro, old Pelileo, old Patate, Sigchos, Angamarca, Chambo (at Guayllabamba bridge), Penipe, Guarando, Ilapo, Guano, San Andrés, Calpi, Cajabamba, Yaruquies (Cacha area), Punín.

15 As interpolated from PRONAREG-ORSTOM, *Isoyetas*.

16 Frank Salomon, *Señores Etnicos*, 99.

Of the 98 square kilometers of non-sandy flats, relic raised fields have been identified as occupying about 20 square kilometers.[17] Most of the flats and the relic fields are in areas experiencing 600–800 mm mean annual precipitation (the notable exceptions being the San Pablo and Paquiestancia raised field areas and Zuleta and Olmedo flats), so that 1,200 persons per square kilometer may have been a reasonable average population density (Table 8.6). Thus anywhere from 24,000 (20 x 1,200) to 117,600 (98 x 1,200) persons could have been supported by raised fields. In all, 74,639 to 168,240 persons (which can be rounded to 75,000 to 170,000 persons) would have occupied the region according to this scenario.

It should be emphasized that this is not a carrying capacity for the region, but rather an estimate of the population which could be supported at a labor investment of 700 person-days per household or less. If households were willing to invest more labor, a greater population could have been supported.

These figures can be checked using other approaches which have already been applied to the task of estimating late pre-Inca population densities.

Historical Demographic Statistics and Depopulation Ratios

Santillán,[18] who resided in Quito between 1563 and 1568 wrote: "there does not remain one fourth of the Indians that there used to be ... in the time of the Inca" (translation mine).[19] This suggests a 4:1 depopulation due to the conflicts and diseases inflicted by the Spanish conquest, but does not include the demographic effects of the earlier Inca conquest.

Paz Ponce de León, writing in 1582, described the Otavalo area:

> They say that in the past there used to be a much greater quantity of Indians ... The Indians have diminished with the wars they waged against the Incas when the latter occupied them, later with the Spanish conquest and finally with certain epidemics which in these parts have been measles, smallpox, and *tabardete* [typhus or typhoid]...[20]

Although there is known to have been a severe epidemic in 1558,[21] what census data there is suggests that population did not change greatly between 1549 and 1575.[22] If there were a 4:1 depopulation it occurred prior to 1549.

---

17 Gondard and López, *Inventario Arqueológico*, 148–149.

18 Hernando de Santillán, "Relación del Origen, Descendencia, Política y Gobierno de los Incas," in Francisco Esteve Barba (ed.), *Crónicas Peruanas de Interés Indígena*, Madrid: Biblioteca de Autores Españoles 209, Ediciones Atlas, 1968, 117.

19 Larrain Barros, *Demografía*, 126.

20 Paz Ponce de León, "Relación," 225. Translation mine.

21 Anonymous, "Cibdad," 205; Salomon, *Señores Etnicos*, 186.

22 Larrain Barros, *Demografía*, Volume 2, 124, 159.

Larrain Barros has calculated (from a remark of Benalcazar's on the number of tributary Indians in the Otavalo chiefdom in 1549) a population of the Otavalo encomienda of 9,400 in 1549. This encomienda apparently included the same area as in 1582, that is the areas from ramp tola sites 11 and 12 on the south, through sites 7, 8, 9, 10, to sites 1, 2, 3, and 4 in the north. In addition, some of this population resided on the western flank of the Andes at Inta(g); it is difficult to know how many, but if half the tributary Indians listed for Cotacachi were actually from Inta in 1598, Inta's proportion of the total population was only 5 percent. If the 4:1 depopulation ratio holds, the Otavalo encomienda may have had roughly 36,000 (4 x 9,000) persons in Inca times.[23]

Another source of information on the Otavalo "chiefdom" population consists of an archival document from 1582 listing the names of the ethnic leaders of the encomienda of Otavalo, specifying the amount of raw cotton which had been distributed to each for the manufacture of textile tribute. Chantal Caillavet has converted the tribute data into population values; her methodology yields a value of about 11,000 persons in the Otavalo chiefdom about 1580.[24] Santillán's depopulation ratio would convert this figure to 44,000 in pre-Hispanic times.

Although by 1582 the population of the Caranque-Cayambe region to the east of Otavalo was under half that of the Otavalo encomienda, there is evidence that this region was especially depopulated by the Inca prior to the Spanish conquest. Thousands of inhabitants of the Cayambe area were deported to other parts of the Inca empire, while Cieza de León reported that at least 20,000 fighting men of the "Carangue" chiefdom had been killed by the Inca army. It therefore seems plausible that both the Caranque (Ibarra) and Cayambe regions were originally more powerful and more populated than the Otavalo encomienda region.[25] If 36,000 to 44,000 persons lived in the area of Otavalo, it may not be excessive to assume that at least an equal number lived in the combined areas of Ibarra (Caranque) and Cayambe in pre-Inca times—or 72,000 to 88,000 persons in the study region shown in Figure 8.1.

Archaeology and Demography

Athens has suggested that all ramp tola sites were occupied simultaneously prior to the Inca conquest, and that the number of ramp tola sites can be multiplied by a standard population value to estimate the population of the region. He suggested the use of a standard average

---

23 Larrain Barros, *Demografia*, Volume 2, 122, 152.

24 Caillavet, "Chefferies Préhispaniques."

25 Larrain Barros, *Demografia*, Volume 2, 124–129, 147; Cieza de León, *Crónica*, 123.

population of 3,000 per tola site, which, if applied to the 22 sites, would result in an estimate of 66,000 persons.[26]

More recent work by Salomon and Caillavet has tended to show the enormous variability in population between parcialidades. About 1580, for example, the various principales of the Otavalo encomienda controlled between 56 and 883 persons each; the two tola sites near San Pablo corresponded to a population of 3,060 controlled by 10 principales, while the four ramp tola sites in the Atuntaqui-Imbaquí-Cotacachi area corresponded to a population of 4,538 under 8 principales. Although the distribution of ramp tola sites seems generally to correspond with the 1580 locations of principales, and easy correlation is not possible.[27]

Nevertheless, if one assigns the 11,000 persons of the 1580 cotton tribute list to 10 ramp tola sites (1–4, 7–10, 11–12), an average of 1,100 persons corresponded with each site; for all 22 sites, a population of 24,200 could be calculated, corresponding (given a 4:1 depopulation) to about 97,000 persons.

Adaptive Dynamics and Populations

Both techniques based on ramp tola distribution and on historic censuses and tribute lists can lead to population estimates consistent with the estimates obtained using the high-intensity scenario of agricultural land use, between 75,000 and 170,000 persons. These results are encouraging in the sense that they do not contradict expectations based on the modelling procedure.

26 Athens, *El Proceso Evolutivo*, 180.

27 Caillavet, "Chefferies Préhispaniques."

Chapter 9

CONCLUSION

The Recent Agricultural Landscape

The present-day agricultural landscape of Ecuador is a diverse mosaic of farms with differing objectives, resources, and technologies. Much of the highlands are still controlled by large farms, which exclude subsistence farmers from access to desirable soils. Ecuadorian price policy and food imports have acted to discourage the commercial production of food crops, and foreign and domestic support has tended to go towards non-agricultural projects favoring animal husbandry or reforestation.[1]

Some large farms, such as the Plaza farms, have persisted in experimenting with food crop production, and their experience provides important evidence for environment-crop interactions. In particular, yields do not necessarily decrease at higher elevations or steeper slopes, maize and potatoes provide comparable amounts of calories at comparable fertilization levels, fertilization can provide great yield improvements even with traditional crops, and excess precipitation is at least as great a hazard as deficient precipitation.

Small farms have produced the bulk of the food crops in highland Ecuador and much of this production is for family consumption. Recent partial land reforms in areas such as the Colta Lake have resulted in initial high yields from fresh soils, decreasing with continued cropping. Hand tools and traditional cropping methods are still widely employed, and their documentation has been one of the main goals of the present book. In particular, soil fertility is managed using household manure supplies, irrigation, fallowing, rotation, and the harvesting of ditch and tank silt. Small farmers have often created diverse and productive landscapes,

---

1 A high agricultural development official in Quito told me in 1987 that highland Ecuador would eventually come to resemble New Zealand, devoted to sheep and dairy cattle raising, and that there was no future for the production of traditional food crops. His agency had virtually stopped funding crop programs in the highlands.

varying from the patchwork of potato and barley fields to the south of Colta Lake to the semi-terraced fields of maize, beans, quinoa, tarwi, barley, and wheat with productive hedgerows and movable corral fertilization systems near Lake San Pablo (Plate 9.1). Especially noteworthy are the numerous canal networks, with their offtakes, tunnels, and tanks, which have been constructed and maintained by traditional practices and which greatly increase the productivity of agriculture in the highlands.

Land use on small farms has often been consistent with the least-effort production of traditional foodstuffs under conditions of a large population on limited land: potatoes are generally restricted to the higher elevations and are managed so as to yield more in calorie terms than maize; very rainy areas are avoided, but canal irrigation is efficient and attractive; most easily irrigable niches are now occupied, often by irrigators organized for the protection of their patrimony; and rotation is practiced, especially for small grains and legumes. Although slope drainage features are constructed and semi-terraces result from slope erosion, bench terraces are rare and raised fields are not constructed or used.

The Late Pre-Inca Equatorial Andes

The attempts to determine some of the characteristics of late pre-Inca agricultural technology and land use allow the drawing of a provisional picture of the equatorial Andes at that time; a picture which is speculative but in line both with evidence and theoretically consistent scenarios.

The picture shows the Andes to have been densely populated, with substantial labor investments in agriculture, the higher elevations used for medium-fallow potato cropping and the lower elevations for short-fallow maize cropping. Nevertheless, at higher elevations, and at very wet and very dry locations at lower elevations, fallows were long enough to permit development of patches of brush and woodland habitat for rabbits, deer, and birds. Soil fertility had often become depleted and the supply of human and guinea pig manure was important for the maintenance of soil fertility.

The most attractive lands were those subject to irrigation, in the sense that the least amount of labor was required to provide for domestic subsistence. These lands were under the control of powerful ethnic lords, and were especially used to produce luxury goods. The long irrigation canals necessary for their use were not onerous to maintain but did provide an element of risk.

Second in labor efficiency, and unsurpassed in security, were the maize lands under 3,100 meters with optimum (900–1,000 millimeters/year) rainfall. These lands had high population densities and were preferred sites both for the general population and for chiefs.

The humid flats had been reclaimed with raised fields, providing drainage, access to water for irrigation, frost control, and, most important, silt and muck accumulation for frequent spreading on field tops. The

Plate 9.1 Contemporary Fields near San Pablo (photo by author)

resulting high soil fertility permitted continuous good yields for double-cropped potatoes and maize. Labor inputs were high, but this drawback was partly offset in the areas of chiefly tola sites due to the presence of an abundant chief's retinue and the availability of corvée labor.

The result was a surprisingly differentiated landscape produced by a single culture, and political order, a diversity within unity which has continued to characterize the Andes to the present, challenging our ability to generalize and predict.

Adaptive Dynamics

This study has resulted in new interpretations of altitudinal and environmental zonation of land use, the strategically relevant environment, and the functions and functional classification of agricultural landforms and strategies. These interpretations are based on explicit specifications of the bases of farmer rationality and choice, which allow for possibilities of testing.

Agriculture has been deliberately dissected into its component techniques in order to focus on particular problems and on the processes of adaptation which help explain farmer choice of one technique over another. This methodology—that of adaptive dynamics—has been chosen in order to study agriculture during a period of time when certain aspects of agriculture were similar but others were different than during the present. Reintegration has been accomplished through the development of simulation models or "scenarios" of labor-minimizing behavior. Although superficially resembling systems models or structural analyses, these scenarios are actually more processual or causal in basing generalizations on explicitly specified factors assumed to influence individual and family choice. The scenarios also specify under what conditions larger-scale organization or identification would be particularly attractive (for example, to manage irrigation) and thus suggest possible pathways toward the understanding of Andean organization.

To repeat a caution made earlier in this study, the purpose of scenarios is to explore the implications of certain assumptions. In the real world, no single factor ever dominates adaptive choice, and no scenario can ever be expected to be fully realized.

If a scenario is clearly contradicted by reality, one may assume that the scenario's assumptions are flawed. The checking of scenario predictions can help identify operative processes, constraints, and selective pressures, but the fit between scenario and reality can always be attributed to coincidence rather than considered a proof that the scenario's assumptions are correct. Differences in interpretation of the significant processes in Andean history can be expected to persist.

This book has been cautious in the employment of modelling and has limited its proposed least-effort explanations to situations where the social, cultural, and demographic context has been independently defined and given due significance. Modelling has not been a substitute for field research into adaptive strategies but rather a tool employed to stimulate and facilitate empirical research.

The affirmation that least-effort models help clarify choice between strategies should not be confused with support for the idea that cultures tend toward energetic efficiency. The present study in fact presents evidence against the latter idea, in that neither prehistoric chiefdoms nor recent nation-states have exhibited labor and land allocations which were optimal for the most efficient provision of calories for subsistence.

Similarly, it is not asserted here that individuals or groups are not capable of assigning cultural value to labor devoted to subsistence purposes, and thus never increase labor beyond the theoretically possible minimum.[2] It is only suggested that all cultures assign greater cultural value to some tasks rather than others, and that the mechanics of food production per se are usually, although not always, seen more as means than goals.

Although it is felt that the operative processes underlying observed patterns have been clarified, the adaptive dynamics approach could be justified on its heuristic value alone: it stimulates the more detailed specification and characterization of adaptive strategies, as historically changeable and contingent on larger demographic, socio-economic, political and cultural processes. In this regard, this book can be seen as part of an evolving tradition of Andean studies, where theoretical differences are ultimately less significant than the common endeavor to learn about an especially fascinating part of the world.[3]

Past Technologies in the Present

This study has especially stressed the power of traditional farmers to modify and enhance soil fertility. In any given adaptive context, some niches will be used more intensively than others, with varying environmental impacts. The variation of land use and human impact is not determined by the "ecology" or inherent "fragility" of land but is rather an historically contingent outcome of a particular way of life: a particular form of production under particular conditions of demand, reflecting a particular shared sense of meaningful goals. fragile lands

---

2 I had the opportunity to interview Billai Laba, a member of a family of raised-field farmers in Papua New Guinea, during his brief visit to Austin, Texas. He mentioned the fact that the worth of men and women in his village was partly judged by the neatness and productivity of their raised fields (which were cultivated by both sexes).

3 For a detailed critique of some of the theoretical approaches presented in this book (and used in earlier studies by the author), see Erickson, *Archaeological Investigation*. The theoretical differences on the salience of labor minimization and Boserup's theory seem to be more apparent than real.

Today the high wetlands of the equatorial Andes are being used as pastures for dairy cattle. The number of Ecuadorians who know of the prehistoric raised fields and their significance is still rather small. Fortunately, initial steps are being taken to preserve some raised fields for posterity. In this way both Ecuadorians and foreign travellers may come to learn of the sophisticated agriculture and dense population of prehistoric Andean peoples. At the same time, people may come to appreciate the fact of agricultural change, and some of the factors involved in the explanation of change. Evidence of change in time and space is the starting point of speculation about origins and diffusions, selective pressures, and processes of choice. In this respect, it is good that the raised fields have been included in the itinerary of geography courses in Ecuador.[4]

The fact that past farming strategies differ from present strategies raises the obvious question of the potential for re-utilization of abandoned technologies. In the Andes, there have been some experiments in the re-utilization of highland raised fields, with encouraging results.[5] In the Ecuadorian case, the existence of highland raised fields has been known for about twenty years. There have apparently been no initiatives to re-establish this technology for small-farm production.[6] In view of the conclusions of this study as to the labor requirements of raised field farming, such initiatives would have to be oriented toward the production of crops of high market value.

Traditional Technologies for the Future

The sophistication of traditional technology should be apparent, as well as its flexibility in dealing with environmental, demographic, socio-economic, and cultural change. If a prerequisite of adaptation is variability, this precondition is amply met in the Andes. Approaches which empower rather than belittle farmers are therefore especially appropriate.

There has been increasing interest in "good farmers" of Latin America, the small farmers using traditional technology, as exemplars for the future.[7]

---

4 In the 1970s and early 1980s, they were at times included in courses offered by the Panamerican Center for Geographical Studies and Investigations (CEPEIGE), under Roy Ryder, Academic Director.

5 Clark L. Erickson and Kay L. Candler, "Raised Fields and Sustainable Agriculture in the Lake Titicaca Basin of Peru," in John O. Browder (ed.), *Fragile Lands of Latin America: Strategies for Sustainable Development*, Boulder, Colorado: Westview Press, 1989, 230–248. Other experiments have occurred in lowland contexts.

6 The re-establishment of raised field farming in highland Ecuador would involve difficult political issues, as much of the good flat land is still in the hands of haciendas.

7 Gene Wilken, *Good Farmers*, Berkeley: University of California Press, 1987.

Invoking the hypotheses that traditional farmers have learned over time to deal with the opportunities and risks of their diverse environments,[8] and have successfully provided for their subsistence in these fragile lands,[9] many now propose that development initiatives which ignore traditional farming technology have little chance for success. There is in fact little evidence that externally introduced technical improvements can do much to increase the labor efficiency of traditional production, at least without subjecting farmers to greater dependency on the cash economy for inputs.

The current Ecuadorian highland situation where farmers seek off-farm cash-earning opportunities and use the income to buy supplemental food and small luxuries has the advantage of not exposing farmers to the risks associated with debt. Until the overall structural context of small-farm agriculture changes, development aid might most appropriately be directed away from agriculture to improving the overall well being and adaptive options of small farmers, through the provision of better health, educational, and transport services (including better roads). Attempts to directly change the agricultural technology probably will not often meet with success, unless subsidized; and such subsidies would be discriminatory and dangerous in that the induced behavior would last only as long as the subsidy persisted.[10]

The successful large scale deployment of agricultural options from the traditional tool kit would have to be coupled, as Zimmerer has pointed out, with a simultaneous favorable context in terms of regional social structures, extra-regional economies, and state agricultural policy. The expansion of wetland agriculture in Coquepata, Peru has involved just such a favorable context.[11] Such a development of favorable contexts, however, is not beyond the ability of local indigenous small-farmers to influence.[12]

Paradoxically, the assertion that traditional decision-makers have been concerned with efficient means to reach social goals does not lead to support for the necessity of development expert intervention, nor does it reduce Andean history as a whole to a unilinear and predictable process. Rather, an approach which focuses on adaptive processes opens vistas of

---

8 David L. Browman (ed.), *Arid Land Use Strategies and Risk Management in the Andes*, Boulder, Colorado: Westview Press, 1987.

9 John O. Browder (ed.), *Fragile Lands of Latin America*, Boulder, Colorado: Westview Press, 1989.

10 Gregory Knapp and Luis Cañadas, "Conclusions and Implications for Policies of Rural Development," in Martin L. Parry, T. R. Carter and N. T. Konijn (eds.), *The Impact of Climatic Variations on Agriculture, Volume 2, Assessments in Semi-Arid Regions*, Dordrecht: Kluwer Academic Publishers, 1988, 485–488.

11 Zimmerer, "Rural Society."

12 Kjell I. Enge and Scott Whiteford, *The Keepers of Water and Earth: Mexican Rural Social Organization and Irrigation*, Austin: University of Texas Press, 1989. This book documents the ability of indigenous farmers in Tehuacán to organize themselves and influence the state, while constructing traditional irrigation systems.

the Andean world as multiple, polyvalent, partially predictable in detail but unpredictable in many of the characteristics which matter most to Andean people. Contemplating the diversity of Andean lifeways one is struck most by elements of surprise and novelty, and the underlying unquenchable creativity (and freedom)[13] which account for this variety.

13 J. C. Scott, *Weapons of the Weak: Everyday Forms of Peasant Resistance*, New Haven: Yale University Press, 1985.

Bibliography

Acosta, Joaquín. *Compendio Histórico del Descubrimiento y Colonización de la Nueva Granada en el Siglo Décimo Sexto*. Paris: Imprenta de Beau, 1848.

Aguilar, Fr. Gerónimo de O.M. "Relación fecha por mi, Fray Gerónimo de Aguilar ... de la Doctrina y Pueblo de Caguasquí y Quilca...[1582]." In Marcos Jiménez de la Espada (ed.), *Relaciones Geográficas de Indias - Perú, Volume 2*. Madrid: Biblioteca de Autores Españoles 184, Ediciones Atlas, 1965, 245-247.

AHBC/I (Archivo Histórico del Banco Central, Ibarra). ["...yo don Marcos Velasques indio principal y alcalde ordinario de pueblo de Carangue..."] (1592). Notarios, Libro 1º, 18 ff.

———— ["...Don Maseo Tulcamaca Cacique Principal del Pueblo de Tulcan..."] (1612). Notarios, Libro 2º, f. 60 r.

———— "Comision en Forma a Luis de Castilejo para que Vaya al Pueblo de Carangue Jurisdicción de la Villa de Ibarra y Restituya a Andres Dias Francisco Ulcoango y Otros Yndios del Pueblo de Carangue de las Tierras que Han Sido Despoxados..." (1635). Juicios, Paquete s/n (Transcribed by Galo Ramón Valarezo).

———— "Testamento de Don Felipe de Gusmán Cazique Principal del Pueblo de Salinas" (1639). Juicios, Paquete No. 10 (1645-1683), 24 ff.

———— "Repartición de Agua de Ambuqui y Repartimiento de las Aguas" (1661). Juicios, Paquete 6, 12 ff.

———— [Sobre las Aguas de la Acequia de Tababuela] (1803). Juicios, Paquete 222.

———— [Litigation between Señor José María Pérez Calisto and Señor Jijón concerning "la acequia sacada de los citios de sus haciendas de Peguchi i Quinchuqui"] (1858). Juicios, Paquete 259 (1846-1858).

———— [Map on folio 22 of incomplete document] (1860?). Juicios, Paquete 209 (1860-1865).

———— [Document of sale of Hacienda La Victoria to Sr. Miguel Teodoro Muñóz] (1874). Notario, 2º, ff. 148,149.

———— "Revindicación de las aguas donadas por la familia Gómez de la Torre..." (1905). Juzgado 1º, No. 94, 202 ff.

AHSA/Q (Archivo Histórico de San Agustín, Quito). "Libro Mayor Títulos y Fondos de este Convento de Nuestro Señor San Agustín, Volumen VIII" (No Date).

Allan, Nigel, Gregory Knapp and Christoph Stadel. (eds.) *Human Impact on Mountains*. Totowa, New Jersey: Rowman and Littlefield, 1988.

Alland Jr., A. "Adaptation." *Annual Review of Anthropology*, Vol. 4 (1975), 59-73.

Anonymous. *Cacique Urcuquí vs. Gamonalismo*. Quito: Editorial Quito, 1946.

Anonymous. "La Cibdad de Sant Francisco de Quito." In Marcos Jiménez de la Espada (ed.), *Relaciones Geográficas de Indias - Perú, Volume 2*. Madrid: Biblioteca de Autores Españoles, Ediciones Atlas, 1965, 205-232.

Antonini, Gustavo, et al. "Chambo: Evolución del Paisaje y su Relación con el Potencial Productivo Agrícola." *Revista del Centro Panamericano de Estudios e Investigaciones Geográficas (Quito)*, Vol. 7 (1981).

Arkin, Herbert and Raymond Colton. *Statistical Methods*. New York: Barnes and Noble, 1970.

Athens, John Stephen. "Evolutionary Process in Complex Societies and the Late Period - Cara Occupation of Northern Highland Ecuador." Ph.D. Dissertation, University of New Mexico, Albuquerque. Ann Arbor: UMI, 1978.

———— *El Proceso Evolutivo en las Sociedades Complejas y La Ocupación del Período Tardío - Cara en los Andes Septentrionales del Ecuador*. Otavalo: Instituto Otavaleño de Antropología, 1980.

Atienza, Lope de. "Compendio Historial del Estado de los Indios del Perú" [1575?]. In Jacinto Jijón y Caamaño (ed.), *La Religión del Imperio de los Incas*. Quito: Escuela Tipográfica Salesiana, 1931, Apéndices, Vol. 1.

Azevedo, J. and P.R. Stout. *Farm and Animal Manures: An Overview of Their Role in the Agricultural Environment*. Berkeley: California Agricultural Experiment Station Manual 44, 1974.

Baker, Paul and Michael Little. (eds.) *Man in the Andes: A Multidisciplinary Study of the High-Altitude Quechua*. Stroudsburg, Pennsylvania: Dowden, Hutchinson, and Ross, 1976.

Bargatzky, Thomas. "Culture, Environment, and the Ills of Adaptationism." *Current Anthropology*, Vol. 25 (1984), 399-406.

Barlett, Peggy F. "Labor Efficiency and the Mechanisms of Agricultural Evolution." *Journal of Anthropological Research*, Vol. 32 (1976), 124-140.

Barsky, Osvaldo and Gustavo Cosse. *Tecnología y Cambio Social: Las Haciendas Lecheras del Ecuador*. Quito: Facultad Latinoamericana de Ciencias Sociales, 1981.

Barth, Fredrik. "Ecologic Relationships of Ethnic Groups in Swat, North Pakistan." *American Anthropologist*, Vol. 58 (1956), 1079-1089.

Basile, David. *Tillers of the Andes: Farmers and Farming in the Quito Basin*. Chapel Hill: University of North Carolina Department of Geography Studies in Geography 8, 1974.

Batchelor, Bruce. "Los Camellones de Cayambe en la Sierra de Ecuador." *América Indígena*, Vol. 40 (1980), 671-689.

Beals, Ralph. *Community in Transition: Nayón-Ecuador*. Los Angeles: University of California at Los Angeles, Latin American Studies Volume 2, 1966.

Bear, Firman E. *Soils in Relation to Crop Growth*. New York: Reinhold, 1965.

Bejarano, Washington José Lainez and Sam Portch. "Adequate Use of Fertilizers on Perennial and Annual Crops in Ecuador." In Elemer Bornemisza and Alfredo Alvarado (eds.), *Soil Management in Tropical America*. Raleigh: North Carolina State University, 1975, 502-517.

Bello Gayoso, Antonio. "Relación que Enbio a Mandar Su Magestad Se Hiziese Desta Ciudad de Cuenca y Toda Su Provincia" [no date]. In Marcos Jiménez de la Espada (ed.), *Relaciones Geográficas de Indias - Perú, Volume 2*. Madrid: Biblioteca de Autores Españoles 184, Ediciones Atlas, 1965, 265-290.

Bennett, John W. "Anticipation, Adaptation, and the Concept of Culture in Anthropology." *Science*, Vol. 192 (1976), 847-853.

———— *The Ecological Transition: Cultural Anthropology and Human Adaptation*. New York: Pergamon, 1976.

———— *Northern Plainsmen: Adaptive Strategy and Agrarian Life*. Chicago: Aldine, 1969.

Blakie, P. and H. C. Brookfield. (eds.) *Land Degradation and Society*. London: Methuen, 1987.

Borja, Fr. Antonio de. "Relación en Suma de la Doctrina e Beneficio de Pimampiro" [1582]. In Marcos Jiménez de Espada (ed.), *Relaciones Geográficas de Indias - Perú, Volume 2*. Madrid: Biblioteca de Autores Españoles 184, Editorial Atlas, 1965, 249-253.

Boserup, Ester. *The Conditions of Agricultural Growth.* Chicago: Aldine, 1965.

Brady, Nyle C. *Nature and Properties of Soils.* New York: Macmillan, 1974.

Brass, L. J. "Stone Age Agriculture in New Guinea." *Geographical Review*, Vol. 31 (1941), 555-569.

Bray, Roger H. "Correlation of Soil Tests with Crop Response to Added Fertilizers and with Fertilizer Requirement." In H.B. Kitchen (ed.), *Diagnostic Techniques for Soils and Crops*. Washington D.C.: American Potash Institute, 1948, 53-86.

Bray, Warwick, Leonor Herrera and Marianne Schrimpff. "Report on the 1981 Field Season in Calima." *Pro Calima (Basel)*, Vol. 3 (1983), 2-31.

Broadbent, Sylvia. "Agricultural Terraces in Chibcha Territory, Colombia." *American Antiquity*, Vol. 29 (1964), 501-504.

__________ "A Prehistoric Field System in Chibcha Territory, Colombia." *Ñawpa Pacha*, Vol. 6 (1968), 135-147.

Brookfield, H. C. "Intensification and Disintensification in Pacific Agriculture: A Theoretical Approach." *Pacific Viewpoint*, Vol. 13 (1972), 30-48.

Brookfield, H. C. and Paula Brown. *Struggle for Land: Agriculture and Group Territories among the Chimbu of the New Guinea Highlands*. Cambridge: Cambridge University Press, 1963.

Browder, John O. (ed.) *Fragile Lands of Latin America.* Boulder, Colorado: Westview Press, 1989.

Browman, David L. (ed.) *Arid Land Use Strategies and Risk Management in the Andes*. Boulder, Colorado: Westview Press, 1987.

Brush, Stephen. "Subsistence Strategies and Vertical Ecology in an Andean Community: Ushucmarca, Peru." Ph.D. Dissertation, University of Wisconsin at Madison, 1973.

__________ "Man's Use of Andean Ecosystem." *Human Ecology*, Vol. 4 (1976), 147-166.

__________ *Mountain, Field and Family: the Economy and Human Ecology of an Andean Valley*. Philadelphia: University of Pennsylvania Press, 1977.

Butzer, Karl W. *Early Hydraulic Civilization in Egypt: A Study in Cultural Ecology*. Chicago: University of Chicago Press, 1976.

__________ *Archaeology as Human Ecology: Theory and Method for a Contextual Approach*. New York: Cambridge University Press, 1982.

__________ "The Realm of Nature and Society: Adaptation and Change in Historical Perspective." In B. L. Turner II et al. (eds.), *The Earth Transformed by Human Action*. New York: Cambridge University Press (Forthcoming).

Caillavet, Chantal. "Etnohistoria Ecuatoriana: Nuevos Datos Sobre el Otavalo Prehispánico." *Cultura (Quito)*, Vol. 5(11) (1981), 109-127.

__________ "Caciques de Otavalo en el Siglo XVI: Don Alonso Maldonado y su Esposa." *Miceldnea Antropológica Ecuatoriana*, Vol. Year 2 (Number 2) (1982), 38-55.

__________ "Toponimia Histórica, Arqueología, y Formas Prehispánicas de Agricultura en la Región de Otavalo - Ecuador." *Boletín del Instituto Francés de Estudios Andinos (Lima)*, Vol. 12 (1983), 1-21.

__________ "La Adaptación de la Dominación Incaica a las Sociedades Autóctonas de la Frontera Septentrional del Imperio (Territorio Otavalo-Ecuador)." *Revista Andina*, Vol. 3 (1985), 403-423.

__________ "La Estructura Básica de las Sociedades Autóctonas de la Sierra Norte del Ecuador en el S. XVI." Paper presented at the Conference on Reproduction and Transformation of Andean Societies of the Social Science Research Council, Quito, June 28-30, 1986.

__________ "Les Chefferies Préhispaniques du Nord de L'Equateur: Formes D'habitat et Organisation Territoriale." *Bulletin Institut Francés Etudes Andines*, Vol. 18 (2) (1988), 41-59.

Caldas, Francisco José de. *Bosquejos Inéditos*. Bogotá: Seminario del Nuevo Reino de Granada, Volume 3, Re-edited by Biblioteca Popular de Cultura Colombiana, Historia 9, Editorial Kelly, 1942.

Cañadas, Luis. *El Mapa Bioclimático y Ecológico del Ecuador*. Quito: Banco Central del Ecuador, 1983.

Cañadas, Luis, Tarquino Salcedo, Raul Bravo and Gregory Knapp. "The Vulnerability of Indigenous Farming Systems to Climatic Variations." In Martin L. Parry, T. R. Carter and N. T. Konijn (eds.), *The Impact of*

Climatic Variations on Agriculture. Volume 2: Assessments in Semi-Arid Regions. Dordrecht: Kluwer Academic Publishers, 1988, 413-428.

Cañadas, Luis and H. Salvador. *Agrometeorological Assessment Models for Rural Development in the Central Sierra of Ecuador*. Columbia, Missouri: Center for Environmental Assessment Services, 1982.

Cañadas, Luis, Trajano Yugcha and Washington Estrada. "The Effects on Zones of Agroclimatic Suitability for Crops." In M.L. Parry, T. Carter and N. T. Konijn (eds.), *The Impact of Climatic Variations on Agriculture. Volume 2: Assessments in Semi-Arid Regions*. Dordrecht: Kluwer Academic Publishers, 1988, 461-471.

Carlstein, Tommy. *Time Resources, Society, and Ecology*. London: George Allen & Unwin, 1982.

CEDIG (Centro Ecuatoriano de Investigación Geográfica). *Documentos de Investigación No. 6, La Erosión en el Ecuador*. Quito, 1986.

CIDA (Comité Interamericano de Desarollo Agricola). *Tenencia de la Tierra y Desarollo Socio-Económico del Sector Agrícola, Ecuador*. Washington D.C.: Panamerican Union, Organization of American States, 1965.

Cieza de León, Pedro de. *La Crónica del Perú [1553]*. Madrid: Espasa-Calpe, 1962.

Clapperton, Chalmers and Ramón Vera. "La Secuencia Glacial del Cuaternario en el Ecuador, una Interpretación al Modelo de W. Sauer." *Paisajes Geográficos (Quito)*, Vol. 16 (1986), 3-20.

Colmet Daage, F. et al. *Caracteristiques de Quelques Sols D'Equateur Derivés de Cendres Volcaniques*. Quito: Programa Nacional de Regionalización Agrícola (Mimeographed), 1968.

Costales, Alfredo and Piedad Peñaherrera de Costales. *Katekil o Historia Cultural del Campesinado del Chimborazo. Llajta No. 4*. Quito: Instituto Ecuatoriano de Geografía e Historia, 1957.

———— *Historia Social del Ecuador, Llajta 17-19*. Quito: Instituto Ecuatoriano de Geografía e Historia, Three Volumes, 1964.

Darch, Janet. (ed.) *Drained Field Agriculture in Central and South America*. Oxford: BAR International Series 189, 1983.

De la Torre, Carlos and Manuel Burga, (eds.), *Andenes y Camellones en el Peru Andino*. Lima: Consejo Nacional de Ciencia y Tecnología, 1986.

Denevan, William M. "Hydraulic Agriculture in the American Tropics: Forms, Measures, and Recent Research." In Kent Flannery (ed.), *Maya Subsistence*. New York: Academic Press, 1982, 181-203.

———— "Adaptation, Variation, and Cultural Geography." *Professional Geographer*, Vol. 35 (1983), 399-407.

———— (ed.) *The Cultural Ecology, Archaeology, and History of Terracing and Terrace Abandonment in the Colca Valley of Southern Peru*. Madison: Technical Report to the National Science Foundation and National Geographic Society (Photocopy), 1986.

Denevan, William M., Kent Mathewson and Gregory Knapp. (eds.) *Pre-Hispanic Agricultural Fields in the Andean Region*. Oxford: BAR International Series 359, 1987.

Dollfus, Olivier. *El Reto del Espacio Andino*. Lima: Instituto de Estudios Peruanos, 1981.

Donkin, Robin. *Agricultural Terracing in the Aboriginal New World*. Tucson: Viking Fund Publications in Anthropology, 1979.

Donohue, Roy, J. Schickluna and L. Robertson. *Soils: An Introduction to Soils and Plant Growth*. Englewood Cliffs, New Jersey: Prentice-Hall, 1971.

Doolittle, William. "Agricultural Change as an Incremental Process." *Annals of the Association of American Geographers*, Vol. 74 (1984), 124-137.

Dunne, Thomas and Luna B. Leopold. *Water in Environmental Planning*. San Francisco: Freeman, 1978.

Echeverría, José. *Contribución al Conocimiento Arqueológico de la Provincia de Pichincha: Sitios Chilibulo y Chillogallo*. Quito: Facultad de Ciencias de la Educación, Pontífica Universidad Católica del Ecuador (Thesis), 1976.

———— *Proyecto: Sistemas de Irrigación Antiguos en la Sierra Norte del Ecuador*. Otavalo: Instituto Otavaleño de Antropología (Manuscript in possession of the author), 1986.

Eidt, Robert. "Aboriginal Chibcha Settlements in Colombia." *Annals of the Association of American Geographers*, Vol. 49 (1959), 374-392.

———— "Detection and Examination of Anthrosols by Phosphate Analysis." *Science*, Vol. 197 (1977), 1327-1333.

__________ "Rural Society and Land Use Change in the Highland Basins of Colombia." *Latin American Studies (University of Tsukuba, Ibaraki Japan)*, Vol. 3 (1981), 25-45.

__________ *Advances in Abandoned Settlement Analysis: Application to Prehistoric Anthrosols in Colombia, South America.* Milwaukee: University of Wisconsin at Milwaukee, The Center for Latin America, 1984.

Elwert, Georg and Diana Wong. "Subsistence Production and Commodity Production in the Third World." *Review (Binghamton, New York)*, Vol. 3 (1980), 501-522.

Enge, Kjell I. and Scott Whiteford. *The Keepers of Water and Earth: Mexican Rural Social Organization and Irrigation.* Austin: University of Texas Press, 1989.

Erasmus, Charles J. "Monument Building: Some Field Experiments." *Southwestern Journal of Anthropology*, Vol. 21 (1965), 277-301.

Erickson, Clark. "Applications of Prehistoric Andean Technology: Experiments in Raised Field Agriculture, Huatta, Lake Titicaca, 1982-1983." In Ian Farrington (ed.), *Prehistoric Intensive Agriculture in the Tropics*. Oxford: B.A.R. International Series 232, 1985, 209-232.

__________ "An Archaeological Investigation of Raised Field Agriculture in the Lake Titicaca Basin of Peru." Ph.D. Dissertation, University of Illinois at Urbana-Champaign. Ann Arbor: UMI, 1988.

Erickson, Clark L. and Kay L. Candler. "Raised Fields and Sustainable Agriculture in the Lake Titicaca Basin of Peru." In John O. Browder (ed.), *Fragile Lands of Latin America: Strategies for Sustainable Development*. Boulder, Colorado: Westview Press, 1989, 230-248.

Farrington, Ian. (ed.) *Prehistoric Intensive Agriculture in the Tropics.* Oxford: BAR International Series 232, 1985.

Ferdon, Edwin, Malcolm Bissell and William Steere. *Studies in Ecuadorian Geography*. Santa Fe, New Mexico: School of American Research Monograph Number 15, 1950.

Field, Leonard and Manuel Chiriboga. *Agricultura Andina.* Quito: CAAP (Centro Andino de Acción Popular), 1984.

Fonseca Martel, César. "Destruccíon de Andenes de la Cuenca del Río Cañete." In Carlos de la Torre and Manuel Burga (eds.), *Andenes y Camellones en el Perú Andino*. Lima: Consejo Nacional de Ciencia y Tecnología, 1986, 361-368.

Fonseca Martel, César and Enrique Mayer. *Comunidad y Producción.* Lima: FOMCIENCIAS, 1988.

Frei, Erwin. *Informe al Gobierno del Ecuador sobre Reconocimientos Edafológicos Exploratorios.* Rome and Quito: Food and Agriculture Organization (Mimeographed), 1957.

Frère, M. J. Rea and J.Q. Rijks. *Estudio Agroclimatológico de la Zona Andina.* Rome: Food and Agriculture Organization, 1975.

Gade, Daniel. "Plant Use and Folk Agriculture in the Vilcanota Valley of Peru: A Cultural-Historical Geography of Plant Resources." Ph.D. Dissertation, University of Wisconsin at Madison, 1967.

―――――― *Plants, Man and Land in the Vilcanota Valley of Peru.* The Hague: W. Junk, 1975.

Garcés, Jorge. (ed.) *Libro Primero de Cabildos de Quito [1529-1538].* Quito: Archivo Municipal de Quito, Publicaciones No. 1, 1934.

―――――― *Libro Primero de la Villa de San Miguel de Ibarra.* Quito: Talleres Tipográficos Municipales, Publicaciones del Archivo Municipal Vol. XV, 1937.

―――――― *Libro de Proveimientos de Tierras, Cuadras, Solares, Aguas, etc., por los Cabildos de la Ciudad de Quito 1583-1594.* Quito: Archivo Municipal Publication XVIII, 1941.

Geertz, Clifford. *Agricultural Involution: The Process of Ecological Change in Indonesia.* Berkeley: University of California Press, 1963.

Glick, Thomas. *Irrigation and Society in Medieval Valencia.* Cambridge: Harvard University Press, 1970.

Glover, J. "Relationship Between Total Seasonal Rainfall and Yield of Maize in the Kenya Highlands." *Journal of Agricultural Science*, Vol. 49 (1957), 285-290.

Golte, Jürgen. *La Racionalidad de la Organización Andina.* Lima: Instituto de Estudios Peruanos, 1987.

Gondard, Pierre. *Inventario y Cartografía del Uso Actual del Suelo en los Andes Ecuatorianos.* Quito: MAG, PRONAREG, ORSTOM, and Museo del Banco Central, 1984.

Gondard, Pierre and Freddy López. *Inventario Arqueológico Preliminar de los Andes Septentrionales del Ecuador.* Quito: MAG-PRONAREG-ORSTOM-Museo del Banco Central, 1983.

Grossman, Lawrence. *Peasants, Subsistence Ecology, and Development in the Highlands of Papua New Guinea.* Princeton: Princeton University Press, 1984.

Guillet, David. "Terracing and Irrigation in the Peruvian Highlands." *Current Anthropology*, Vol. 28 (1987), 409-430.

Hall, Minard. *El Volcanismo en el Ecuador*. Quito: Instituto Panamericano de Geografía e Historia, 1977.

Harris, Marvin. *Cows, Pigs, Wars and Witches: the Riddles of Culture.* New York: Random House, 1974.

Hastenrath, Stefan. *The Glaciation of the Ecuadorian Andes*. Rotterdam: A.A. Balkema, 1981.

Hatch, John K. *The Corn Farmers of Motupe: A Study of Traditional Farming Practices in Northern Coastal Peru.* Madison: University of Wisconsin, Land Tenure Center, Monographs Number 1, 1976.

Heider, Karl G. *The Dugum Dani: A Papuan Culture in the Highlands of West New Guinea.* New York,: Viking Fund Publications in Anthropology 49, Wenner-Gren Foundation for Anthropological Research, Inc., 1970.

Holdridge, Leslie R. *Life Zone Ecology*. San José, Costa Rica: Tropical Science Center, 1967.

Humboldt, Alexander and Aimé Bonpland. *Essai sur la géographie des plantes; accompagné d'un tableau physique des régions equinoxiales.* Mexico City: Institut Panamericain de Géographie et d'Histoire, Editorial Cultura, 1955.

INAMHI (Instituto Nacional de Meteorología e Hidrología). *Anuario Meteorológico, 1981*. Quito, 1983.

INEC-MAG (Instituto Nacional de Estadística y Censos - Ministerio de Agricultura). *Sistema Estadística Agropecuario Nacional por Muestreo por Areas,* Quito, 1985.

INERHI/I (Instituto Ecuatoriano de Recursos Hidráulicos, Ibarra). "Concession I-73-589: Acequia 'Grande o de Caciques'" (1973-1982).

INERHI/Q (Instituto Ecuatoriano de Recursos Hidráulicos, Quito). "Proyecto Esmeraldas - Unidad Técnica, Inventario de los Usos de Agua en la Cuenca del Río Monjas: Cuadros Catastrales y Mapas de Ubicación." (1973).

__________ "Proyecto Esmeraldas - Unidad Técnica, Inventario de los Usos de Agua en las Cuencas de los Ríos Chiche, Guambi, Uravia, y Coyago." (1974).

———— "Inventario de los Usos de Agua, Subcuenca del Rio Ambi" (1977).

———— "Inventario de los Usos de Agua, Subcuenca del Rio Ambi" (1977).

———— "Inventario del Uso de Agua, Ríos Palacara, Huarmiyacu, Cariyacu, y Pingunchela" (1980). .

INIAP (Instituto Nacional de Investigaciones Agropecuarias). *Programa de Investigación en Producción, PIP Imbabura: Informe Annual 1983*. Quito, 1983.

———— *Programa de Investigación en Producción, PIP Imbabura: Ciclo Agrícola 1984-1985*. Quito, No Date.

Jaramillo, Victor Alejandro. *Corregidores de Otavalo*. Otavalo: Instituto Otavaleño de Antropología, 1972.

Jijón y Caamaño, Jacinto. *Contribución al Conocimiento de los Aborígenes de la Provincia de Imbabura en la República del Ecuador*. Madrid: Blass y Cía., Estudios de la Prehistoria Americana 2, 1914.

Jochim, Michael. "Breaking Down the System: Recent Ecological Approaches in Archaeology." *Advances in Archaeological Method and Theory*, Vol. 2 (1979), 77-116.

———— *Strategies for Survival: Cultural Behavior in an Ecological Context*. New York: Academic Press, 1981.

Johnson, A.M. "The Climate of Peru, Bolivia, and Ecuador." In W. Schwerdtfeger (ed.), *Climates of Central and South America*. Amsterdam: Elsevier, 1976, 147-218.

Keller et al., Jack. *Ecuador Irrigation Sector Review. USAID/Ecuador Water Management Synthesis Project Report 12*. Logan: Agricultural and Irrigation Engineering, Utah State University, 1982.

Kirch, Patrick. "The Archaeological Study of Adaptation: Theoretical and Methodological Issues." In Michael Schiffer (ed.), *Advances in Archaeological Method and Theory*. New York: Academic Press, 1980, Vol. 3, 101-156.

Knapp, Gregory. *The Sunken Fields of Chilca: Horticulture, Microenvironment and History in the Peruvian Coastal Desert*. Madison: Unpublished M.S. thesis, Department of Geography, University of Wisconsin, 1979.

———— "El Nicho Ecológico Llanura Húmeda, en la Economía Prehistórica de los Andes de Altura." *Sarance*, Vol. 9 (1981), 83-96.

―――――――― "Cambios en el Clima de los Andes Ecuatorianos: Una Tentativa Explicación Genética." In Roy Ryder and Nelson Robelly (eds.), *Geografía y Desarollo: Metodologías y Casos de Estudio*. Quito: Instituto Geográfico Militar, 1982, 19-26.

―――――――― "Prehistoric Flood Management on the Peruvian Coast: Reinterpreting the 'Sunken Fields' of Chilca." *American Antiquity*, Vol. 47 (1982), 144-154.

―――――――― "Soil, Slope, and Water in the Equatorial Andes: A Study of Prehistoric Agricultural Adaptation." Ph.D. Dissertation, University of Wisconsin at Madison. Ann Arbor: UMI, 1984.

―――――――― "Riego Precolonial en la Sierra Norte." *Ecuador Debate (Quito)*, Vol. 14 (1987), 17-45.

―――――――― *Ecología Cultural Prehispánica del Ecuador*. Quito: Banco Central del Ecuador, 1988.

―――――――― "The Effects of Variations in Mean Temperatures and Frost Risk." In Martin L. Parry, T. R. Carter and N. T. Konijn (eds.), *The Impact of Climatic Variations on Agriculture. Volume 2: Assessments in Semi-Arid Regions*. Dordrecht: Kluwer Academic Publishers, 1988, 443-460.

Knapp, Gregory and Luis Cañadas. "Conclusions and Implications for Policies of Rural Development." In Martin L. Parry, T. R. Carter and N. T. Konijn (ed.), *The Impact of Climatic Variations on Agriculture, Volume 2, Assessments in Semi-Arid Regions*. Dordrecht: Kluwer Academic Publishers, 1988, 485-488.

Knapp, Gregory and William Denevan. "The Use of Wetlands in the Prehistoric Economy of the Northern Ecuadorian Highlands." In Ian S. Farrington (ed.), *Prehistoric Intensive Agriculture in the Tropics*. Oxford: BAR International Series, 1985, 185-207.

Knapp, Gregory and David Preston. "Evidence of Prehistoric Ditched Fields on Sloping Land in Northern Highland Ecuador." In William Denevan, Kent Matthewson and Gregory Knapp (eds.), *The Ecology and Archaeology of Prehispanic Agricultural Fields in the Central Andes*. Oxford: BAR International Series 359, 1987, 403-424.

Knapp, Gregory and Roy Ryder. "Aspects of the Origin, Morphology, and Function of Ridged Fields in the Quito Altiplano, Ecuador." In Janet P. Darch (ed.), *Drained Field Agriculture in Central and South America*. Oxford: B.A.R. International Series 189, 1983, 201-220.

Kolata, Alan. "The Agricultural Foundations of the Tiwanaku State: A View from the Heartland." *American Antiquity*, Vol. 51 (1986), 748-762.

———— *La Tecnología y Organización de la Producción Agrícola en el Estado de Tiwanaku: Primer Informe de Resultados del Proyecto Wilajawira*. La Paz: Editorial Sui Generis, 1989.

Kolata, Alan and Charles Ortloff. "Thermal Analysis of Tiwanaku Raised Field Systems in the Lake Titicaca Basin of Bolivia." *Journal of Archaeological Science*, Vol. 16 (1989), 233-263.

Kus, James. "Selected Aspects of Irrigated Agriculture in the Chimú Heartland." Ph.D. Dissertation, University of California at Los Angeles. Ann Arbor: UMI, 1972.

Larrain Barros, Horacio. *Cronistas de Raigambre Indígena*. Otavalo: Instituto Otavaleño de Antropología (Two Volumes), 1980.

———— *Demografía y Asentamientos Indígenas en la Sierra Norte del Ecuador en el Siglo XVI*. Otavalo: Instituto Otavaleño de Antropología (Two Volumes), 1980.

Lathrap, Donald, John Isaacson and Colin McEwan. "On the Trail of the Finest Metallurgy of the Ancient New World: How Old is the Classic Quimbaya Style?" *Field Museum of Natural History Bulletin*, Vol. 55(10) (1984), 10-19.

Lechtman, Heather and Ana Maria Soldi, (eds.), *La Tecnología en el Mundo Andino*. Mexico City: Universidad Nacional Autónoma de Mexico, 1981.

Leung, Woot-Tsuen Wu. *Tabla de Composición de Alimentos para Uso en América Latina*. Mexico City: Interamericana, 1964.

Lippi, Ronald. "Paleotopography and Phosphate Analysis of a Buried Jungle Site in Ecuador." *Journal of Field Archaeology*, Vol. 15 (1988), 85-97.

MAG-INERHI (Ministerio de Agricultura - Instituto Ecuatoriano de Recursos Hidraulicos). *Tomas de Estudio de Factibilidad, Proyecto Tabacundo*. Quito: (7 volumes), 1978.

Masuda, Shozo Izumi Shimada and Craig Morris. (eds.) *Andean Ecology and Civilization*. Tokyo: University of Tokyo Press, 1985.

Mathewson, Kent. "Taxonomy of Raised and Drained Fields: A Morphogenetic Approach." In Ian Farrington (ed.), *Prehistoric Intensive Agriculture in the Tropics*. Oxford: BAR International Series 232, 1985, 835-845.

Mayer, Enrique. *Land Use in the Andes: Ecology and Agriculture in the Mantaro Valley of Peru with Special Reference to Potatoes*. Lima: Centro Internacional de la Papa, 1979.

__________ "Production Zones." In Shozo Masuda, Izumi Shimada and Craig Morris (ed.), *Andean Ecology and Civilization*. Tokyo: University of Tokyo Press, 1985, 45-84.

Meyers, Albert. "Análisis de la Cerámica de Cochasquí." In Udo Oberem (ed.), *Cochasquí: Estudios Arqueológicos*. Otavalo: Instituto Otavaleño de Antropología, 1981, 219-285.

Mitchell, William. "Irrigation and Community in the Central Peruvian Highlands." *American Anthropologist*, Vol. 78 (1976), 25-44.

Molestina, María del Carmen. "Investigaciones Arqueológicas en la Zona Negativo del Carchi o Capulí." *Cultura (Quito)*, Vol. 7 (21a) (1985), 31-82.

Monroy, Joel O. M. "El Convento de la Merced de Quito de 1534 a 1617 (1a Parte)." *Boletín de la Academia Nacional de Historia (Quito)*, Vol. 10 (1930), 137-184.

Moran, Emilio. (ed.) *The Ecosystem Concept in Anthropology*. Boulder: Westview Press, 1984.

Mothes, Patricia. *Pimampiro's Canal: Adaptation and Infrastructure in Northern Ecuador*. Austin: Unpublished M.A. thesis, Department of Geography, University of Texas at Austin, 1986.

__________ "La Acequia del Pimampiro: Riego Tradicional en el Norte del Ecuador." *Ecuador Debate*, Vol. 14 (1987), 69-86.

Murra, John. "The Historic Tribes of Ecuador." In Julian Steward (ed.), *Handbook of South American Indians, Volume* 2. Washington, D.C.: Smithsonian Institution Bureau of American Ethnology, Bulletin 143, 1944, 785-822.

__________ "Rite and Crop in the Inca State." In S. Diamond (ed.), *Culture and History: Essays in Honor of Paul Radin*. New York: Columbia University Press, 1960, 393-407.

__________ "El 'Control Vertical' de un Máximo de Pisos Ecológicos en la Economía de las Sociedades Andinas." In John Murra (ed.), *Visita de la Provincia de León de Huánuco en 1562*. Huánuco, Peru: Universidad Nacional Hermilio Valdizán, 1972, 429-476.

Myers, Thomas. "Evidence of Prehistoric Irrigation in Northern Ecuador." *Journal of Field Archaeology*, Vol. 1 (1974), 309-313.

Nabas, Juan de Dios. *Ibarra y Sus Provincias de 1534 a 1932*. Quito: Imprenta del Clero (2 Volumes), 1934.

NRC (Commission on International Relations National Resource Council Board on Science and Technology for International Development). *Postharvest Food Losses in Developing Countries,* Washington D.C.: National Academy of Sciences, 1978.

Odum, Eugene P. *Fundamentals of Ecology*. Philadelphia: Saunders, 1953.

Olsen, S.R. and L.A. Dean. "Phosphorus." In C. A. Black (ed.), *Methods of Soil Analysis*. Madison: American Society of Agronomy, 1965, 1035-1049.

Ordóñez de Cevallos, Pedro. "Historia y Viaje del Mundo del Padre Agradecido [1614]." (ed.), *Cronistas Coloniales 2*. Puebla, Mexico: Biblioteca Ecuatoriana Mínima, Editorial Cajica Jr., 1960, 475-519.

Orlove, Benjamin. "Integration through Production: The Use of Zonation in Espinar." *American Ethnologist,* Vol. 4 (1977), 84-101.

———— "Ecological Anthropology." *Annual Review of Anthropology,* Vol. 9 (1980), 235-273.

Parry, Martin. "A Typology of Cultivation Ridges in Southern Scotland." *Tools and Tillage,* Vol. 3 (1976), 3-19.

Parry, Martin and Tim Carter. *Assessing the Impact of Climate Change in Cold Regions*. Laxenburg: International Institute for Applied Systems Analysis, 1984.

Parsons, James. "The Northern Andean Environment." *Mountain Research and Development,* Vol. 2 (1982), 253-262.

Paz Maldonado, Fr. Juan. "Relación del Pueblo de Sant-Andrés Xunxi para el muy ilustre Señor Licensiado Francisco de Auncibay, del Consejo de su Majestad y su Oydor en la Real Audiencia de Quito [1582]." In Marcos Jiménez de la Espada (ed.), *Relaciones Geográficas de Indias - Perú, Volume 2*. Madrid: Biblioteca de Autores Españoles 184, Ediciones Atlas, 1965, 261-264.

Paz Ponce de León, Sancho de. "Relación y Descripción de Los Pueblos del Partido de Otavalo [1582]." In Marcos Jiménez de la Espada (ed.), *Relaciones Geográficas de Indias - Perú, Volume 2*. Madrid: Biblioteca de Autores Españoles, Ediciones Atlas, 1965, 233-242.

Plaza Schuller, Fernando. *Contribución al Estudio de los Montículos Prehistóricos de los Andes Septentrionales del Ecuador: Apuntes de*

Aerofotointerpretación Arqueológica. Otavalo: Instituto Otavaleño de Antropología (Manuscript Atlas), 1977.

Porras, Pedro I. *Fase Cosanga.* Quito: Ediciones de la Universidad Católica, 1975.

Pourrut, Pierre. "Los Climas del Ecuador: Fundamentos Explicativos." (ed.), *Los Climas del Ecuador.* Quito: Centro Ecuatoriano de Investigación Geográfica, Documentos de Investigación No. 4, 1983, 7-41.

Preston, David A. *Field Lines in Northern Highland Ecuador: Preliminary Account of Field Observations.* Leeds: University of Leeds School of Geography, Working Paper 380, Revised Version, 1984.

PRONAREG-ORSTOM (Programa Nacional de Regionalización Agraria - Office de la Recherche Scientifique et Technique Outre-Mer). *Estudio Hidro-Meteorológico e Hidrogeológico Preliminar de las Cuencas de los Ríos Esmeraldas y del Norte Ecuatoriano,* Quito, 1977.

__________ *Diagnóstico Socio-Económico del Medio Rural Ecuatoriano, Documentos 1-11.* Quito: Ministerio de Agricultura - ORSTOM, 1978.

__________ *Ecuador: Mapa Ecológico,* 1:1,000,000 Map, Quito: Instituto Geográfico Militar, 1978.

__________ *Diagnóstico Socio-Económico del Medio Rural Ecuatoriano, Documentos A-D.* Quito: Ministerio de Agricultura - ORSTOM, 1979.

__________ *Isoyetas Medias Anuales e Histogramas Mensuales, Período 1964-1973,* 1:1,000,000 Map, Quito: Instituto Geográfico Militar, 1979.

__________ *Mapas de Suelos por Regionalización,* 1:50,000 Maps, Martinique and Quito: Centre des Antilles - Bureau des Sols, 1979.

__________ *El Agua con Fines de Riego. Provincias de Carchi e Imbabura,* Quito: Ministerio de Agricultura - ORSTOM, 1982.

__________ *Mapas de Uso Actual del Suelo y Formaciones Vegetales,* 1:200,000 Maps, Quito: Instituto Geográfico Militar, 1983.

__________ *Los Principales Procesos Erosivos en Ecuador,* Quito, 1984.

Pulgar Vidal, Javier. *Historia y geografía del Perú: Tomo I, Las ocho regiones naturales del Perú.* Lima: Universidad Nacional Mayor de San Marcos, 1946.

Ramón Valarezo, Galo. "Del Cacicazgo Andino a la Hacienda: La Transformación del Espacio Productivo en Cayambe." *Cultura (Quito)*, Vol. 8 (24b) (1986), 639-654.

Rappaport, Roy. *Pigs for the Ancestors: Ritual in the Ecology of a New Guinea People*. New Haven: Yale University Press, 1967.

Ravelo, Andrés, Ana M. Planchuelo-Ravelo and Tom Hodges. "The Effects of Climatic Variations and Agricultural Practices on Barley Yields." In M. L. Parry, T. Carter and N. T. Konijn (eds.), *The Impact of Climatic Variations on Agriculture. Volume 2: Assessments in Semi-Arid Regions*. Dordrecht, Netherlands: Kluwer Academic Publishers, 1988, 429-442.

Ravines, Rogger. (ed.) *Tecnología Andina*. Lima: Instituto de Estudios Peruanos, 1978.

Reichel-Dolmatoff, Gerardo. *Colombia*. New York: Frederick Praeger, 1965.

Reinhold, Brian. "Weather Regimes: The Challenge in Extended Range Forecasting." *Science*, Vol. 235 (1987), 437-441.

Riley, Thomas and Glen Freimuth. "Field Systems and Frost Drainage in the Prehistoric Agriculture of the Upper Midwest." *American Antiquity*, Vol. 44 (1979), 271-285.

Rodríguez de Aguayo, Pedro. "Descripción de la Ciudad de Quito y vecindad de ella por el arcediano de su iglesia, licensiado Pedro Rodríguez de Aguayo" [No Date]. In Marcos Jiménez de la Espada (ed.), *Relaciones Geográficas de Indias - Perú, Volume 2*. Madrid: Biblioteca de Autores Españoles 184, Ediciones Atlas, 1965, 201-204.

Rodríguez Docampo, Diego. "Descripción y Relación del Estado Eclesiástico del Obispado de San Francisco de Quito que se Ha Hecho por Mandado del Rey Nuestro Señor en Virtud de su Real Cédula Dirigida al Illmo. Sr. D. Agustín Ugarte Saravia..." [1650]. In Marcos Jiménez de la Espada (ed.), *Relaciones Geográficas de Indias - Perú, Volume 3*,. Madrid: Biblioteca de Autores Españoles 184, Ediciones Atlas, 1965, 3-78.

Rovere, Oscar and Gregory Knapp. "Selection of Climatic Scenarios." In Martin L. Parry, T. R. Carter and N. T. Konijn (eds.), *The Impact of Climatic Variations on Agriculture*. Dordrecht: Kluwer Academic Publishers, 1988, 399-412.

Rubio Orbe, Gonzalo. *Punyaro*. Quito: Casa de la Cultura Ecuatoriana, 1956.

Ryder, Roy. "El Valor de la Fotografía Aérea en los Estudios Históricos y Arqueológicos del Ecuador." *Revista Geográfica del IGM (Quito)*, Vol. 6 (1970), 40-42.

Sahlins, Marshall. *Culture and Practical Reason*. Chicago: University of Chicago Press, 1976.

Salomon, Frank. *Los Señores Etnicos de Quito en la Epoca de los Incas*. Otavalo, Ecuador: Instituto Otavaleño de Antropología, 1980.

——————— *Native Lords of Quito in the Age of the Incas: The Political Economy of North Andean Chiefdoms*. Cambridge: Cambridge University Press, 1986.

Sanchez, Pedro, Walter Couto and Stanley Buol. "The Fertility Capability Soil Classification System: Interpretation, Applicability, and Modification." *Geoderma*, Vol. 27 (1982), 283-309.

Sánchez, Rodrigo. "The Model of Verticality in the Andean Economy: A Critical Reconsideration." *Actes du XLIIe Congrès International Des Americanistes*, Paris, (1976), Vol. IV, 213-231.

Santillán, Hernando de. "Relación del Origen, Descendencia, Política y Gobierno de los Incas." In Francisco Esteve Barba (ed.), *Crónicas Peruanas de Interés Indígena*. Madrid: Biblioteca de Autores Españoles 209, Ediciones Atlas, 1968, 97-150.

Sauer, Carl O. "Cultivated Plants of South and Central America." In Julian Steward (ed.), *Handbook of South American Indians*. Washington: Smithsonian Institution Bureau of American Ethnology Bulletin 143, 1950, Volume 6, 487-543.

Sauer, Carl O. and Donald Brand. "Prehistoric Settlements of Sonora, with Special Reference to Cerros de Trincheras." *University of California Publications in Geography*, Vol. 5 (6) (1931), 67-148.

Sauer, Walther. *Geología del Ecuador*. Quito: Ministerio de Educación, 1965.

Schroeder, Kathleen. "Señora Córdova's Kitchen: Food Substitution in the Sierra of Piura." Paper presented at the Annual Meeting of the Association of American Geographers, Toronto, Canada, April 19-22, 1990.

Scott, J.C. *Weapons of the Weak: Everyday Forms of Peasant Resistance*. New Haven: Yale University Press, 1985.

SGM (Servicio Geográfico Militar). *Guaillabamba, (Provincia de Pichincha)*, 1:25,000 Map, Quito: Servicio Geográfico Militar, No Date (c. 1940?)

Sherbondy, Jeanette. "The Canal Systems of Hanan Cuzco." Ph.D. Dissertation, University of Illinois at Urbana-Champaign, Ann Arbor: UMI, 1982.

———— (ed.) *Antiguedad y Actualidad del Riego en los Andes*. Cusco: Instituto de Pastoral Andina, Allpanchis Numbers 26 and 27, 1986.

Smith, Clifford T., William M. Denevan and Patrick Hamilton. "Ancient Ridged Fields in the Region of Lake Titicaca." *The Geographical Journal*, Vol. 134 (1968), 353-367.

Stadel, Christoph. "Environmental Stress and Human Activities in the Tropical Andes (Ecuador)." *Revista del Centro Panamericano de Estudios e Investigaciones Geográficas (Quito)*, Vol. 15 (1985), 33-50.

Steensberg, Axel. *New Guinea Gardens: A Study of Husbandry with Parallels in Prehistoric Europe*. New York: Academic Press, 1980.

Steward, Julian. "Ecological Aspects of Southwestern Society." *Anthropos*, Vol. 32 (1937), 87-104.

———— *Theory of Culture Change: The Methodology of Multilinear Evolution*. Urbana: University of Illinois Press, 1955.

Thomas, R. Brooke. "Energy Flow at High Altitude." In Paul Baker and Michael Little (eds.), *Man in the Andes: A Multidisciplinary Study of the High-Altitude Quechua*. Stroudsburg, Pennsylvania: Dowden, Hutchinson, and Ross, 1976, 379-404.

Thompson, Louis and Frederic Troeh. *Soil and Soil Fertility*. New York: McGraw Hill, 1973.

Tobar Subía, Cristobal. *Monografía de Ibarra*. Ibarra: Prensa Católica, 1950.

Tosi Jr., Joseph A. *Zonas de Vida Natural en el Perú*. Lima: Instituto Interamericano de Ciencias Agrícolas de la OEA: Zona Andina, 1960.

Treacy, John. "The Fields of Coporaque: Agricultural Terracing and Water Management in the Colca Valley, Arequipa, Peru." Ph.D. Dissertation, University of Wisconsin, Madison. Ann Arbor: UMI, 1989.

Troll, Carl. "The Cordilleras of the Tropical Americas. Aspects of Climatic, Phytogeographical, and Agrarian Ecology." In Carl Troll (ed.), *Geo-Ecology of the Mountainous Regions of the Tropical Americas*. Bonn: Colloquium Geographicum Band 9, Ferd. Dümmlers Verlag, 1968, 13-56.

Turner II, B. L. and Stephen Brush. (eds.) *Comparative Farming Systems*. New York: The Guilford Press, 1987.

Turner II, B. L. and William Doolittle. "The Concept and Measure of Agricultural Intensity." *Professional Geographer*, Vol. 30 (1978), 297-301.

Uhle, Max. "The Aims and Results of Archaeology [1923]." In John Rowe (ed.), *Max Uhle, 1856-1944*. Berkeley: University of California Publications in Archaeology and Ethnography, 1954, 54-100.

Uribe, Maria Victoria. "Asentamientos Prehispánicos en el Altiplano de Ipiales, Colombia." *Revista Colombiana de Antropología*, Vol. 21 (1977-1978), 57-196.

USDA (United States Department of Agriculture). *Soil Taxonomy*. Washington: U.S. Government Printing Office, 1975.

———— *Annual Agricultural Situation Report, Ecuador*. Quito: (Mimeographed), 1987.

Vargas, María and Sylvia Gallegos. *Diagnóstico Nutricional y Alimentario del Ecuador*. Riobamba: Escuela Superior Politécnica de Chimborazo, No Date.

Velasco, Juan de. *Historia del Reino de Quito en la América Meridional* [1789]. Quito: Editorial Casa de la Cultura Ecuatoriana (3 Volumes), 1977-1979.

Vera, Ramón. "El Origen de Cangahua." *Paisajes Geográficos (Quito)*, Vol. 16 (1986), 21-28.

Waddell, Eric. *The Mound Builders*. Seattle: University of Washington Press, 1972.

Watts, Michael. *Silent Violence: Food, Famine and Peasantry in Northern Nigeria*. Berkeley: University of California Press, 1983.

Webster, Steven. "An Indigenous Quechua Community in Exploitation of Multiple Ecological Zones." *Actas y Memorias del XXXIX Congreso Internacional de Americanistas*, Lima, Vol. 3 (1971), 174-183.

Weismantel, Mary. *Food, Gender and Poverty in the Ecuadorian Andes*. Philadelphia: University of Pennsylvania Press, 1988.

Whitehead, Alfred N. *Process and Reality*. New York: Macmillan, 1929.

Wilken, Gene. *Good Farmers*. Berkeley: University of California Press, 1987.

Willcox, O.W. *ABC of Agrobiology*. New York: Norton, 1937.

Willcox, O.W. "Verification of the Mitscherlich Effect Law." *Agronomy Journal*, Vol. 43 (1949), 83-90.

———— "Quantitative Agrobiologic Evaluation of Some Phosphate Fertilizer Trials with Vegetable Crops in Alabama." *Agronomy Journal*, Vol. 43 (1951), 83-90.

Winterhalder, Bruce. "Environmental Analysis in Human Evolution and Adaptation Research." *Human Ecology*, Vol. 8 (1980), 135-170.

Winterhalder, Bruce, Robert Larsen and R. Brooke Thomas. "Dung as an Essential Resource in a Highland Peruvian Community." *Human Ecology*, Vol. 2 (1974), 89-104.

Wolf, Teodoro. *Geografía y Geología del Ecuador*. Quito: Casa de la Cultura Ecuatoriana, 1975.

Wray Espinosa, Marcos. *Proyecto de Desarollo Ganadero de la Hacienda "Santa Natalia": Pedro Moncayo-Pichincha*. Quito: Facultad de Ingenería Agronómico y Medicina Veterinaria, Universidad Central del Ecuador (Thesis), 1976.

Wurster, Wolfgang. "Aportes a la Reconstrucción de Edificios con Planta Circular, Sobre las Pirámides con Rampa de Cochasquí." In Udo Oberem (ed.), *Cochasquí: Estudios Arqueológicos*. Otavalo: Instituto Otavaleño de Antropología, 1981, 79-124.

Zimmerer, Karl. "Seeds of Peasant Subsistence: Agrarian Structure, Crop Ecology and Quechua Agriculture in Reference to the Loss of Biological Diversity in the Southern Peruvian Andes." Ph.D. Dissertation, University of California at Berkeley. Ann Arbor: UMI, 1988.

———— "Rural Society and Wetland Ecology in the Agricultural Transformation of a Highland Peruvian Region." *Annals of the Association of American Geographers*, (In Press).

Zipf, George Kingsley. *Human Behavior and the Principle of Least Effort: An Introduction to Human Ecology*. Cambridge: Addison-Wesley Press, 1949.

Zumárraga, Pedro M. *Monografía del Cantón Antonio Ante*. Quito: La Prensa Católica, 1949.

Index

www.ingramcontent.com/pod-product-compliance
Ingram Content Group UK Ltd.
Pitfield, Milton Keynes, MK11 3LW, UK
UKHW021502240926
475702UK00035B/176
* 9 7 8 0 3 6 7 0 1 6 3 2 6 *